UMBELLIFERS
OF THE
BRITISH ISLES

B.S.B.I. Handbook No. 2

T. G. TUTIN

*Emeritus Professor of Taxonomy in the
University of Leicester*

ILLUSTRATED BY
ANN DAVIES

BOTANICAL SOCIETY OF THE BRITISH ISLES
London
1980

Published by the Botanical Society of the British Isles
c/o British Museum (Nat. Hist.), Cromwell Road, London, S.W.7
Printed by the Devonshire Press, Torquay, Devon

CONTENTS

ACKNOWLEDGEMENTS

My thanks are due to the members of the Publications Committee of the B.S.B.I. for many useful comments, and especially to Dr C. A. Stace, frequent discussions with whom have done much to improve the manuscript. Dr S. L. Jury and Mr M. J. Southam have also provided valuable information on various aspects of the umbellifers about which I was ignorant. I am greatly indebted to Mr J. F. M. Cannon, not only for comments but also for the general supervision of the illustrations drawn by Mrs Ann Davies, whose skill and patience are so well reflected in the result.

August 1979   T.G.T.

INTRODUCTION

There is a popular impression that the Umbelliferae, with a few conspicuous exceptions such as *Eryngium* and *Bupleurum*, are almost indistinguishable from one another unless you have ripe fruit and a microscope. In fact with a little experience almost all British umbellifers can be identified when in flower and often from the leaves alone. The main objection to using leaf-characters in a book is the difficulty of describing shape, dissection, margin, apex, venation, texture etc. in a clear and unambiguous way. The eye can learn to appreciate and recognize the sum total of these characters while the pen remains baffled. Smell, even less describable, is also often diagnostic.

There are no critical species or groups among British umbellifers and few hybrids occur; *Oenanthe* is the genus most likely to give trouble and it may be impossible to distinguish *O. lachenalii* from *O. silaifolia* until they are in fruit. With this exception no serious difficulty should be encountered.

The obvious distinctive characters of the majority of the Umbelliferae made them one of the earliest plant families to be generally recognized, though foreign elements, such as members of the Valerianaceae, were usually included and *Eryngium* and *Hydrocotyle* excluded. The importance of identifying individual species must have been apparent to primitive man; to mistake *Oenanthe crocata* for celery could have fatal consequences. It was probably this mixture of plants valuable for food or flavouring with others superficially similar but poisonous which led in the first place to accurate identification.

A concise history of the classification of the Umbelliferae is given by Lincoln Constance (1971) and need not be repeated here.

Most members of the family are herbs with hollow or pith-filled flowering stems. *Bupleurum* contains a number of shrubby species, one of which is naturalized in England, and other woody members of the family occur in Africa and South America; among the largest is the African *Steganotaenia* which is a tree up to 12m high.

The cotyledons of the seedlings have been extensively investigated by M.-Th. Cerceau-Larrival who finds that they fall into two classes, those that taper imperceptibly at the base and those which are abruptly contracted into a petiole. As far as is known the type of cotyledon is constant within a genus. Seedlings of perennial and biennial species develop a rosette of leaves, which often show a gradation of form, the earliest to arise being less divided than the later; these rosette leaves are usually rather different from those borne on the flowering stems and it is the latter which are described here, unless rosette-leaves are specifically mentioned. Annuals usually have leaves which are more or less evenly spaced and show the same sort of variation as those on the flowering stems of biennials and perennials. In the majority of species the leaves are pinnately divided, most frequently two or more times, and there is a diminution in size and degree of division passing from the base of the stem upwards. Simply pinnate or palmately lobed leaves are much less frequent and entire ones are found only in *Bupleurum*. Ternate leaves, i.e. those which are divided into three more or less equal parts, each of which may again be similarly divided, also occur not infrequently. The ultimate distinctly stalked divisions of a leaf are referred to as lobes; they may be entire, toothed or pinnatifid (i.e. deeply cut, but not to the midrib).

Leaves are usually petiolate and spirally arranged. The petioles of the lower leaves are often long and have a sheathing base, while the upper leaves usually have short, often entirely sheathing and sometimes greatly inflated petioles. The venation of the leaf is usually pinnate and characteristic, but some species of *Bupleurum* have leaves with parallel veins; these probably represent flattened petioles, rather than leaf-blades. There are no stipules, except in the subfamily Hydrocotyloideae.

The commonest kind of inflorescence is a compound umbel, which consists of a number of branches ('rays') arising at the same level and each bearing several flowers which also all arise at the same level, forming a partial umbel. It has usually been supposed that the umbel has been derived from a racemose inflorescence by the suppression of internodes, but evidence has recently been produced which suggests that in the Hydrocotyloideae derivation from a branched cymose inflorescence (thyrse) is more likely (Froebe, 1971). Simple umbels, presumably the equivalent of one ray of a compound umbel, occur in some members of the subfamilies Hydrocotyloideae and Saniculoideae and in *Hydrocotyle* itself the flowers are often in whorls. In *Eryngium* the flowers are sessile and crowded into thistle-like heads with each individual flower subtended by a bracteole. The umbel terminates a stem or branch, the oldest one being on the

4

main stem and usually having the largest number of rays. Such umbels are referred to as terminal umbels, those on the branches being called lateral umbels. The peduncle may be long, so that the terminal umbel overtops the lateral umbels, or it may be almost non-existent, as in *Torilis nodosa*, where each successive umbel appears to be lateral and leaf-opposed and the stem unbranched.

At the point where the rays arise from the peduncle there is often a number of bracts; the presence or absence of these, their number, size, shape and texture can provide useful characters. Similarly the presence or absence of bracteoles subtending the partial umbel can be of diagnostic value.

The individual flower is small, though the umbel is often large and conspicuous. The five sepals are free and usually smaller than the petals or not infrequently completely absent. The petals alternate with the sepals (if present) and are most often concave, with an inflexed apex, so that they appear to be notched or 2-lobed. The outer petals of flowers at the periphery of the umbel are sometimes much larger than the remainder, when they are said to be radiating; otherwise the flowers are actinomorphic. There is often an oil canal along the middle of the petal, and the petal itself may be hairy or papillose beneath. Umbellifer petals have been largely neglected by taxonomists (except in *Bupleurum*), but would probably provide useful characters if carefully examined.

The stamens alternate with the petals in a single whorl and are inflexed in bud. The filaments vary from much shorter than to longer than the petals in different genera and the anthers are seldom much longer than wide (*Eryngium* is a notable exception) and are attached to the filament at about the middle of the back.

The ovary is inferior and 2-locular, rarely 1-locular through abortion in a few genera unrepresented in Britain; each loculus contains a single pendent ovule and the pericarp has typically five vascular bundles in each carpel. These are usually evident as wings or ribs in the fruit. The ovary has a nectar-secreting disc at its summit from which two styles arise; these often have thickened bases and together with the disc form the stylopodium. The styles vary considerably in length, being extremely short in *Bupleurum* and long and flexuous in *Pimpinella*; they may be erect, spreading, or recurved and appressed to the stylopodium. The stigmas may be unthickened, slightly and gradually thickened, a more or less globose knob, or capitate. ·

The two carpels are joined by a commissure, along which they separate at maturity (except in *Coriandrum*); they generally hang for a time from the apex of the usually bifid carpophore which lies between the two oil canals (vittae) on the commissural face of each carpel. The carpophore and commissural vittae are absent in the Hydrocotyloideae and Saniculoideae. Between each pair of vascular bundles in the pericarp on the outer face of the carpel there is usually a single vitta, rarely absent or split into several small vittae; in a few genera there are secondary ridges which alternate with and may be more prominent than the primary ridges. If these are present the vittae are concealed by them. The carpels may be almost semi-circular in cross-section or compressed to various degrees parallel to or at right angles with the commissure; if the former, they are described as dorsally compressed and if the latter, as laterally compressed. In some genera, notably *Scandix*, the carpels have a distinct beak, a continuation of the carpel beyond the part containing the seed. The surface of the fruit is sometimes hairy, covered with papillae, or with straight or hooked bristles or spines, the latter usually arising from the ridges.

Although the fruit is basically uniform throughout the family, it shows a variety of adaptations for different means of dispersal. In *Crithmum* and some species of *Oenanthe*, for instance, the pericarp is thick and corky and the mericarps float well; dorsally compressed fruits, particularly those with the lateral ridges winged, offer a large expanse to catch the wind, and those with hooked spines easily get tangled in the hair of animals. In general the size of the fruit and its ridges or wings shows little variation within a species, but there are exceptions; a striking example of fruit variability has been described recently in a species of *Cachrys*, where the extremes had for years been regarded as separate species (Herrnstadt & Heyn, 1975).

In many Umbelliferae a proportion of the flowers have shorter styles (sometimes none) and smaller stylopodia than the remainder, and the ovary is abortive so that they are functionally male, while the remainder are hermaphrodite. Male flowers occur particularly frequently in lateral umbels, the latest to develop very often having no hermaphrodite flowers. Most genera are protandrous, a state which presumably encourages outbreeding; the comparatively few genera in which the anthers dehisce and the stigmas become receptive simultaneously must be self-pollinated to a large extent, unless self-incompatibility exists; this has been found to be a rare occurrence in the family as far as it has been investigated. The significance of the

6

varying proportions of hermaphrodite and male flowers appears to be unknown and indeed there is very little information available about it. The only dioecious species of umbellifer in the British Isles is the very local *Trinia glauca* and even here the separation of the sexes is often incomplete.

The umbelliferous flower is apparently unspecialized, usually with fully exposed nectar secreted, often abundantly, by the disc or stylopodium. Insects, mostly small and themselves unspecialized as pollinators, may usually be seen sitting on or walking about any umbel when in flower and several hundred species have been recorded as visitors. Diptera are the commonest, followed by unspecialized Hymenoptera, such as ants and short-tongued bees. There appears to have been no investigation of the floral biology of umbellifers in this country or in the rest of Europe, but preliminary observations of considerable interest have been made in the United States (Bell, 1971), which suggest that similar work on this side of the Atlantic would be rewarding. Bell found evidence of some degree of specialization, for example in those species in which the basal half of the infolded petals is erect and forms a sort of corolla-tube a millimetre or more long which tends to limit access to the nectar to short-tongued Hymenoptera. The shape, size and colour of the stylopodium is probably also of significance. Thrips are abundant on umbellifer flowers and on account of their small size must be able to evade any known restrictive mechanism and carry out a certain amount of self-pollination.

CLASSIFICATION AND RELATIONSHIPS

The Umbelliferae fall fairly clearly into three subfamilies, Hydrocotyloideae, Saniculoideae and Apioideae, the first of which has sometimes been regarded as a separate family. The present tendency is to combine the Umbelliferae with Araliaceae, rather than to split the Umbelliferae. The main differences between these two families are given in the following table.

Umbelliferae	*Araliaceae*
Herbs, rarely woody	Woody, rarely herbs
Ovary 2-locular (rarely 1-locular)	Ovary 1- to many-locular
Endocarp usually not woody	Endocarp usually woody
Fruit very rarely fleshy	Fruit usually fleshy

There is thus no single character which distinguishes the two families, though they can nevertheless nearly always be easily recognized. Chemical evidence also suggests that they are closely related and could well have diverged from a common stock— Araliaceae mainly tropical, woody, with fleshy fruits and the hard endocarp that is so often a feature of pyrenes that have to withstand a bird's digestive system; Umbelliferae mainly temperate, with membranous endocarp and dry fruits adaptable to many means of dispersal. Araliaceae remain morphologically more variable, though nearly always with juicy fruits, while Umbelliferae tend to follow one pattern, with their remarkably uniform flowers and fruits. The highly specialized annual or biennial habit occurs frequently in Umbelliferae. Stellate hairs are common in Araliaceae and also occur, though less frequently, in Umbelliferae. In both families 22 is probably the commonest diploid chromosome number and polyploidy appears to be much more common in Araliaceae than in Umbelliferae.

On the basis of chemical characters Hegnauer suggests that the Umbellales are not a climax group but may be descended from a stock which also gave rise to the Asterales and which in turn was perhaps derived from the Rutalean stock.

The family Umbelliferae has already been described in some

detail (pp. 4–6). Brief diagnoses of the three subfamilies will be given here but no division into tribes etc. will be attempted. The tribal classification most often used is that of Drude (1898), but this is greatly in need of revision to take into account the abundant information now available about cytology, phytochemistry, anatomy and palynology, so no useful purpose would be served by repeating it here.

Subfamily **Hydrocotyloideae**: Leaves often simple, usually with scarious stipules. Flowers usually in simple umbels or in whorls, sometimes solitary. Ovary with a more or less flat disc. Fruit with a woody endocarp and no vittae, at least when mature. Carpophore absent. Basic chromosome number most commonly 8. Widely distributed, mostly in the southern hemisphere.
Only British genus: *Hydrocotyle*.

Subfamily **Saniculoideae**: Leaves often simple or palmately lobed, without stipules. Flowers in simple umbels or capitula. Ovary with a more or less flat disc. Fruit with a membranous endocarp. Vittae usually present in mature fruit. Carpophore absent. Basic chromosome number most commonly 8. Widely distributed.
British genera: *Sanicula*, *Astrantia*, *Eryngium*.

Subfamily **Apioideae**: Leaves usually much divided, without stipules. Flowers commonly in compound umbels. Ovary with a prominent stylopodium. Fruit with a membranous endocarp. Vittae usually present in the mature fruit. Carpophore present. Basic chromosome number most commonly 11. Cosmopolitan, but particularly abundant in the northern hemisphere.

British genera: *Chaerophyllum, Anthriscus, Scandix, *Myrrhis, *Coriandrum, *Smyrnium, Bunium, Conopodium, Pimpinella, *Aegopodium, Sium, Berula, Crithmum, Seseli, Oenanthe, Aethusa, *Foeniculum, *Anethum, Silaum, Meum, Physospermum, Conium, Bupleurum, Trinia, Apium, *Trachyspermum, Petroselinum, Sison, Cicuta, *Ammi, *Falcaria, Carum, Selinum, Ligusticum, Angelica, *Levisticum, Peucedanum, Pastinaca, Heracleum, Tordylium, Torilis, Daucus*.

The genera marked with an asterisk are certainly or probably introduced by human agency.

HABITAT LISTS

Umbellifers grow in a great variety of habitats and are often common in places strongly affected by human activities. They are absent from true salt-marshes, from the upper parts of the higher mountains and from extensive areas of peat on moors, heaths and mountains. The following lists are intended as a guide to the species which can be expected in various habitats, but must not be taken as exhaustive. An asterisk indicates that the species has certainly or probably been introduced by human agency.

Woods
   Sanicula europaea

Wood-margins, clearings, shady hedge-banks
   Angelica sylvestris, Anthriscus sylvestris, *Astrantia major, Chaerophyllum temulentum, Conopodium majus, Heracleum sphondylium, Pimpinella major, Physospermum cornubiense, Torilis japonica.

Dry grassy places
   *Bupleurum fruticosum, Daucus carota, Eryngium campestre, *Falcaria vulgaris, Heracleum sphondylium, Petroselinum segetum, Pimpinella saxifraga, Sison amomum, *Smyrnium perfoliatum, Torilis nodosa.

   on chalk: Bunium bulbocastanum, Pastinaca sativa, Seseli libanotis.

   on limestone: Trinia glauca.

   on acid soils: Conopodium majus.

   in mountain regions: Meum athamanticum, *Peucedanum ostruthium.

Damp grassland and fens
   Angelica sylvestris, Carum verticillatum, Conium maculatum, Hydrocotyle vulgaris, Oenanthe fistulosa, O. lachenalii, O.

pimpinelloides, O. silaifolia, Peucedanum palustre, Selinum carvifolia, Silaum silaus.

Cliffs and clay banks near the sea
Bupleurum baldense, Crithmum maritimum, Daucus carota, *Foeniculum vulgare, Ligusticum scoticum, Peucedanum officinale, *Smyrnium olusatrum, Trinia glauca.

Dunes
Crithmum maritimum, Daucus carota, Eryngium maritimum, Pimpinella saxifraga.

Brackish habitats
Anthriscus caucalis, Apium graveolens, Bupleurum tenuissimum, Conium maculatum, Oenanthe lachenalii, *Smyrnium olusatrum, Torilis nodosa.

In fresh water ditches, ponds, canals and slow-flowing rivers
Apium inundatum, A. nodiflorum, Berula erecta, Cicuta virosa, Oenanthe aquatica, O. crocata, O. fistulosa, O. fluviatilis, O. lachenalii, Sium latifolium.

On river-banks
Angelica archangelica, Conium maculatum, *Heracleum mantegazzianum.

Arable land
Aethusa cynapium, Scandix pecten-veneris, Torilis arvensis, T. nodosa (very rarely *Bupleurum rotundifolium, *Caucalis platycarpos, *Turgenia latifolia).

Roadsides and open hedge-banks
*Aegopodium podagraria, Anthriscus caucalis, A. sylvestris, *Carum carvi, *Chaerophyllum aureum, C. temulentum, Conium maculatum, Conopodium majus, *Foeniculum vulgare, *Heracleum mantegazzianum, H. sphondylium, *Myrrhis odorata, Pastinaca sativa, Petroselinum segetum, Pimpinella major, P. saxifraga, Sison amomum, *Smyrnium olusatrum, Tordylium maximum, Torilis japonica, T. nodosa.

Sphagnum *bogs*
Hydrocotyle vulgaris.

12

Garden weeds
  *Aegopodium podagraria, Aethusa cynapium.

Weeds from bird-seed
  *Bupleurum subovatum, *Trachyspermum ammi.

Cultivated plants that sometimes persist
  *Anethum graveolens, *Levisticum officinale, *Smyrnium perfoliatum.

Lawns
  Conopodium majus, *Hydrocotyle moschata.

Rubbish dumps
  *Aegopodium podagraria, Aethusa cynapium, *Ammi majus, *A. visnaga, *Angelica archangelica, *Bupleurum subovatum, Conium maculatum, *Coriandrum sativum, *Trachyspermum ammi.

A COOK'S GUIDE TO THE UMBELLIFERAE

The culinary heyday of the umbellifers in British kitchens was really the long period before the introduction of the turnip for winter cattle-feed when the bulk of winter meat supplies was salt beef and pork. Many strong-flavoured umbellifers, such as *Myrrhis*, were used to disguise the not very appetizing taste and smell; first with the aid of the turnip and still more with the introduction of refrigeration these have completely gone out of use. However, quite a number of species still find a regular place in the kitchen either as a vegetable or for flavouring or as a garnish, and the current interest in exotic cooking coupled with the influence of the many immigrants who have come to Britain, means that the umbellifers are making a strong domestic come-back. The chief of these are listed below and the part of the plant employed and its main culinary uses are indicated.

Angelica (*Angelica archangelica*). Petiole, young stem and 'seeds'.
Caraway (*Carum carvi*). 'Seeds' and rarely leaves.
Carrot (*Daucus carota* subsp. *sativus*). Root.
Celeriac (*Apium graveolens* var. *rapaceum*). Stock.
Celery (*A. graveolens* var. *dulce*). Petiole and stem.
Chervil (*Anthriscus cerefolium*). Leaf.
Coriander (*Coriandrum sativum*). 'Seeds'.
Dill (*Anethum graveolens*). Leaf and 'seeds'.
Fennel (*Foeniculum vulgare* subsp. *vulgare*). Leaf.
Finocchio (*F. vulgare* subsp. *vulgare* var. *azoricum*). Enlarged leaf-bases.
Lovage (*Levisticum officinale*). Leaf.
Parsley (*Petroselinum crispum*). Leaf.
Parsnip (*Pastinaca sativa* subsp. *sativa*). Root.

Chief uses
　Soup: Caraway (leaf), celery, chervil, dill, fennel, lovage.
　Sauces: Angelica, chervil, dill ('seed'), fennel, parsley (chiefly with fish or vegetables); lovage (with poultry).
　Salads: Angelica (petiole), carrot (grated), chervil, lovage.

Meat: Caraway ('seeds') with goulash, coriander ('seeds') with curried lamb.

Puddings: Carrots (grated).

Bread and cakes: Angelica (candied petiole), caraway ('seeds').

Pickles (flavouring): Dill (leaf and 'seed').

Vegetables: Carrot, celery, celeriac, finocchio, parsnip.

Garnishes: Fennel, lovage, parsley.

Liqueurs (flavouring): Angelica ('seed'), caraway ('seed').

Pimpinella anisum, aniseed, is used for flavouring fancy bread and cakes, a liqueur (Anise) and a sweet (aniseed balls), but does not appear to be grown in the British Isles. The tubers of *Conopodium majus* (pignuts) are eaten by children and *Aegopodium podagraria* is said to be used in soups and stews in C. & E. Europe (see Hegi, 1926, and Komarov, 1950). It was probably introduced to this country for these purposes, and as a green vegetable, but is no longer used.

Crithmum, Myrrhis and *Smyrnium olusatrum* also appear to have gone out of use, together with a number of other more local species.

KEYS

Explanation

Two distinct kinds of key for the identification of the British genera and species of Umbelliferae are presented here. The first is the familiar dichotomous key, in which each entry is numbered at the left hand side of the page and the two entries having the same number contrast with one another. Having decided which of the alternatives presented in the first entry applies most nearly to your plant you may, for example, find that you have arrived at *Eryngium*, in which case you continue with the key to species, where the alternatives are indicated by letters. If the plant is not an *Eryngium* proceed to the entry numbered 2 and continue the process until you arrive at an entry which fits your plant and has the name of a species opposite to it on the right hand side of the page. In genera with more than one species the numbered part of the key will take you to a generic name and the subsequent key to the individual species in that genus has the entries indicated by letters, instead of figures. The entries are indented throughout to make it easier to pick out contrasting pairs.

A rudimentary form of this kind of key was the 'hierarchical tree' presented by Robert Morison in his work on Umbelliferae published in 1672. Morison may in fact have been the inventor of this valuable aid to identification, as no earlier example appears to be known (Hedge, 1973), though he did not suggest that it should be used as we now use keys.

The second kind of key, first constructed by Hedge and Lamond (1964), is known as a Multi-Access Key. Here a table of alternative states (in this case 20) is provided, each being given a letter (A to T). The plant to be identified is compared with this table and the nine letters denoting the various characters which it displays are written down. The resulting assemblage of letters or formula is then looked up in the alphabetical list of formulae, where it may occasionally be found to refer to a single species. More often a number of different species have the same formula, and then a series of distinctive features, which should enable the species or genus to be identified, is given. Where two or more species of the same genus key out together a separate key to the species has been given (keys A–O). It is hoped

16

that these will also be useful to the beginner when he has sufficient experience to know at a glance that he is dealing with a *Bupleurum* or a *Torilis*. He can then go straight to the key to the species of that genus, rather than toil through the main key. The multi-access key will be found particularly useful when the material is incomplete and some of the characters therefore missing. The various possibilities offered by the incomplete 'formula' can be found from the alphabetical list and a choice between them made by comparing the plants with the descriptions.

An example may help to explain the method of using keys of this kind. A plant to be identified has the following characters: flowers not yellow (B), lowest leaves at least 2-pinnate (E), fruit less than 3 times as long as wide (G) and without hairs, spines or papillae (I), no fibres surrounding the base of the stem (K), umbel compound and bracteoles present (M), perennial (P), no deflexed, appressed hairs on the stem (R), and unwinged fruit (S). Thus the formula BEGIKMPRS is obtained, and on referring to the alphabetical list it will be found that three genera and five other species have this formula. The statement 'Bracts all 3-fid or pinnatisect' means that of the plants with this formula *Ammi* alone possesses bracts of this kind. Similar additional characters are given where they are needed to distinguish different genera or species which have the same formula or where they are particularly useful in confirming the identification.

Whichever key is used it is important to compare the plant with description and illustration in order to confirm the identification.

DICHOTOMOUS KEY

1 Leaves coriaceous, with spiny margins; flowers in dense heads
ERYNGIUM

  (a) Leaves 3(–5)-lobed; flowers and often the stems bluish
4. E. maritimum

  (a) Leaves pinnate; flowers whitish; stems pale green
5. E. campestre

1 Leaves usually herbaceous, without spines; flowers in umbels or rarely whorls

  2 All leaves entire BUPLEURUM

    (a) Upper cauline leaves perfoliate; bracts absent

      (b) Fruit not papillose   **38. B. rotundifolium**

      (b) Fruit papillose   **39. B. subovatum**

    (a) Upper cauline leaves not perfoliate; bracts present, at least at flowering time

      (c) Perennial

        (d) Herb; leaves parallel-veined   **42. B. falcatum**

        (d) Shrub; leaves with a distinct midrib and network of lateral veins   **43. B . fruticosum**

      (c) Annual

        (e) Leaves parallel-veined, without distinct cross-veins; bracteoles ovate, concealing the flowers **40. B. baldense**

        (e) Leaves parallel-veined, with distinct cross-veins; bracteoles subulate, not concealing the flowers
41. B. tenuissimum

  2 Leaves toothed or more deeply divided

    3 Upper cauline leaves perfoliate, ovate, denticulate or crenulate
15. Smyrnium perfoliatum

    3 Upper cauline leaves not perfoliate

      4 Leaves ± orbicular, divided for less than half-way; stems very slender, creeping HYDROCOTYLE

        (a) Leaves peltate, glabrous   **1. H. vulgare**

        (a) Leaves with a deep sinus, usually hairy
H. moschata (see **1**)

      4 Leaves divided for more than half-way; stems not very slender and creeping

        5 Basal leaves palmately divided

6 Flowers subsessile in an irregular compound umbel; involucre inconspicuous **2. Sanicula europaea**

6 Flowers with long filiform pedicels, in a simple umbel surrounded by a conspicuous, pink or white involucre
3. Astrantia major

5 Basal leaves pinnately or ternately divided

  7 Plant aquatic with finely divided, translucent submerged leaves and few or no aerial leaves at flowering time

    8 Umbel with 1–2(–4) rays; sepals minute or absent
48. Apium inundatum

    8 Umbel with (4–)5–16 rays; sepals relatively conspicuous   OENANTHE

      (a) Fruit 3.5–4.5 mm, ovoid **30. O. aquatica**

      (a) Fruit 5–6.5 mm, cylindrical **29. O. fluviatilis**

  7 Plant usually terrestrial with aerial leaves numerous and well-developed at flowering time; submerged leaves rarely present

    9 Lowest aerial leaves pinnately lobed to simply pinnate (c.f. p. 21)

      10 Plant hairy (use lens)

        11 Sepals conspicuous, about equalling petals

          12 Umbel with more than 4 rays
69. Tordylium maximum

          12 Umbel with 2–4 rays
Turgenia latifolia (see **72**)

        11 Sepals minute or absent

          13 Flowers yellow **66. Pastinaca sativa**

          13 Flowers white or pink

            14 Bracteoles absent   PIMPINELLA

              (a) Stem hollow, prominently angled; lower leaves with 3–4 pairs of lobes
17. P. major

              (a) Stem solid, at least when young, terete; lower leaves with (2–)4–7 pairs of lobes
18. P. saxifraga

            14 Bracteoles several

              15 Hairs on stem not deflexed and appressed; fruit winged   HERACLEUM

              (a) Umbel with 10–20 rays
67. H. sphondylium

(a) Umbel with 50–150 rays
68. H. mantegazzianum

15 Hairs on stem deflexed and closely appressed; fruit spiny or tuberculate
TORILIS

(a) Umbels subsessile, leaf-opposed
70. T. nodosa

(a) Umbels long-pedunculate
(b) Bracts 0–1; fruit 4–6 mm
71. T. arvensis

(b) Bracts 4–6(–12); fruit 3–3.5 mm
72. T. japonica

10 Plant glabrous
16 Petiole fistular; partial umbels subglobose in fruit   **24. Oenanthe fistulosa**

16 Petiole not fistular; partial umbels not subglobose in fruit
17 Lower leaves with more than 20 pairs of lobes   **57. Carum verticillatum**

17 Lower leaves with fewer than 10 pairs of lobes
18 Longest bracts at least $\frac{1}{2}$ as long as the shortest rays
19 Bracts subulate; stem solid
51. Petroselinum segetum

19 Bracts wider, often leaf-like; stem hollow
20 Umbel usually with 20–30 rays; petals papillose beneath; fruit 3–4 mm
20. Sium latifolium

20 Umbel usually with 7–14 rays; petals smooth beneath; fruit 1.5–2 mm
21. Berula erecta

18 Bracts absent or all less than $\frac{1}{2}$ as long as the rays
21 Some umbels subsessile and leaf-opposed   APIUM

(a) Bracteoles absent
45. A. graveolens

(a) Bracteoles 3–7
(b) Stem rooting at lower nodes only;

peduncle usually shorter than rays;
bracts 0–2 **46. A. nodiflorum**
(b) Stem rooting at all nodes;
peduncle usually longer than rays;
bracts 3–7 **47. A. repens**
21 Umbels all with long peduncles, not leaf-opposed
22 Bracts 2–4 **52. Sison amomum**
22 Bracts absent
23 Bracteoles present
12. Coriandrum sativum
23 Bracteoles absent
24 Leaves ternate; long rhizomes present
19. Aegopodium podagraria
24 Leaves pinnate; plant without rhizomes PIMPINELLA
(a) Stem hollow, prominently angled; lower leaves with 3–4 pairs of lobes **17. P. major**
(a) Stem solid, at least when young, terete; lower leaves with (2–)4–7 pairs of lobes
18. P. saxifraga
9 Lower leaves at least 2-pinnate or -ternate
25 Flowers yellow or greenish-yellow
26 Bracts more than 3; bracteoles numerous
27 Bracteoles connate below
62. Levisticum officinale
27 Bracteoles free to the base
28 Leaf-lobes subulate to linear-lanceolate
22. Crithmum maritimum
28 Leaf-lobes ovate
29 Lobes of lower leaves crenate or serrate, obtuse **13. Smyrnium olusatrum**
29 Lobes of lower leaves pinnatifid, acute
50. Petroselinum crispum
26 Bracts 0–3; bracteoles usually few or none
30 Leaf-lobes filiform, entire
31 Annual **33. Anethum graveolens**
31 Perennial

32 Stock without fibrous remains of petioles; bracteoles absent
   32. Foeniculum vulgare

32 Stock with fibrous remains of petioles; bracteoles present
   63. Peucedanum officinale

30 Leaf-lobes toothed or lobed, not linear

 33 Lobes of lower leaves crenate or serrate, obtuse **13. Smyrnium olusatrum**

 33 Lobes of lower leaves pinnatifid, acute

  34 Sheathing petiole-base without or with a narrow hyaline margin; leaf-margin serrulate **34. Silaum silaus**

  34 Sheathing petiole-base with a broad hyaline margin; leaf-margin entire
   50. Petroselinum crispum

25 Flowers white, greenish-white or pinkish

 35 Bracts all 3-fid or pinnatisect

  36 Biennial, ± hispid; central flower of umbel usually purple: fruit spiny
   73. Daucus carota

  36 Annual, glabrous; central flower of umbel white; fruit not spiny AMMI

   (a) Rays slender and spreading after flowering **54. A. majus**

   (a) Rays thickened and contracted after flowering **A. visnaga** (see **54**)

 35 Bracts absent or mostly undivided

  37 Plant with a tuber from which arises 1 stem, which tapers downwards and is flexuous from soil level to its junction with the tuber

  38 Stem hollow after flowering; styles suberect in fruit **16. Conopodium majus**

  38 Stem always solid; styles recurved in fruit
   15. Bunium bulbocastanum

  37 Flowering stems arising at or near soil level, not tapering and flexuous downwards; root-tubers sometimes present

  39 Stems conspicuously swollen below the nodes

40 Rays 3–5 **Anthriscus cerefolium** (see **9**)
40 Rays (4–)8–25 CHAEROPHYLLUM
  (a) Biennial; leaves dark green; fruit 5–6 mm; styles as long as the stylopodium **6. C. temulentum**
  (a) Perennial; leaves yellowish; fruit c. 9 mm; styles much longer than the stylopodium **7. C. aureum**
39 Stems not conspicuously swollen below the nodes
  41 Annual, with slender roots; lower leaves usually dead at flowering time
    42 Bracteoles 2– to 3–fid or pinnatifid; fruit with a long linear beak which develops soon after flowering **10. Scandix pecten-veneris**
    42 Bracteoles entire; fruit without a long beak
      43 Fruit densely grey-papillose **49. Trachyspermum ammi**
      43 Fruit not densely grey-papillose
        44 Sepals conspicuous, often as long as the petals
          45 Stems glabrous; fruit smooth **12. Coriandrum sativum**
          45 Stems with deflexed, appressed hairs; fruit spiny **Caucalis platycarpos** (see **72**)
        44 Sepals small or absent
          46 Stems glabrous or with sparse, patent or forward-pointing hairs
            47 Bracteoles deflexed, on the outer side of the partial umbels only **31. Aethusa cynapium**
            47 Bracteoles not deflexed, on all sides of the partial umbels **9. Anthriscus caucalis**
          46 Stems with deflexed, appressed hairs, at least above TORILIS
            (a) Umbels subsessile, leaf-opposed **70. T. nodosa**
            (a) Umbels long-pedunculate
              (b) Bracts 0–1; fruit 4–6 mm **71. T. arvensis**
              (b) Bracts 4–6(–12); fruit 3–3.5 mm **72. T. japonica**
  41 Biennial or perennial with stout stock, rhizome or tap-root; lower leaves often green at flowering time
    48 Base of stem surrounded by abundant fibrous remains of petioles

49 Usually dioecious; male plants with dense umbels and numerous shortly pedicellate flowers; female plants with lax umbels and few long-pedicellate flowers
44. Trinia glauca

49 Most flowers hermaphrodite; umbels all similar
50 Leaf-lobes lanceolate to ovate in outline
23. Seseli libanotis
50 Leaf-lobes filiform　　　**35. Meum athamanticum**
48 Base of stem without fibrous remains of petioles
51 Basal and lower cauline leaves 1–to 3– ternate
52 Leaf-lobes linear to linear-lanceolate, the midrib (at least near the base) with a vein on either side close to and parallel with it　　**55. Falcaria vulgaris**
52 Leaf-lobes rhombic in outline, often 3-fid at apex, the midrib without accompanying parallel veins
53 Leaves puberulent on margin and veins
36. Physospermum cornubiense
53 Leaves glabrous　　　**59. Ligusticum scoticum**
51 Basal and lower cauline leaves pinnately divided
54 Bracteoles on outer side of partial umbels only; ovary and fruit usually with crenulate-undulate ridges　　　**37. Conium maculatum**
54 Bracteoles absent or on all sides of the partial umbels; ridges on ovary and fruit not crenulate-undulate
55 Rays puberulent or papillose, at least on the angles (use lens)
56 Stems and leaves ± hairy
57 Sheaths of upper leaves not greatly inflated; plant strongly aromatic
11. Myrrhis odorata
57 Sheaths of upper leaves greatly inflated; plant not strongly aromatic　　ANGELICA
(a) Flowers white or pinkish; stem usually purplish; fruit 4–5 mm　**60. A. sylvestris**
(a) Flowers greenish; stem usually green; fruit c. 6 mm　**61. A. archangelica**
56 Stems and leaves glabrous
58 Bracts present, persistent, deflexed
64. Peucedanum palustre
58 Bracts absent or few, caducous, not deflexed

59 Leaf-lobes 5–10 cm; rays usually more
than 30   **65. Peucedanum ostruthium**

59 Leaf-lobes 0.3–1 cm; rays usually fewer
than 25   **58. Selinum carvifolia**

55 Rays glabrous and smooth
  60 Bracteoles absent (rarely 1)
    61 Umbels with peduncle shorter than rays
        45. Apium graveolens

    61 Umbels with peduncle longer than rays
      62 Umbels mostly with fewer than 10 rays;
      fruit smelling of caraway when crushed
        56. Carum carvi

      62 Umbels with more than 10 rays; fruit
      not smelling of caraway when crushed
        63 Rhizome far-creeping; plant glabrous
        19. Aegopodium podagraria

        63 Rhizome absent; plant with at least
        some small crispate hairs   PIMPINELLA
          (a) Stem hollow, prominently angled;
          lower leaves with 3–4 pairs of lobes
        17. P. major

          (a) Stem solid, at least when young,
          terete; lower leaves with (2–) 4–7
          pairs of lobes   **18. P. saxifraga**

  60 Bracteoles present
    64 Plant hairy, the hairs sometimes small and
    crispate; sepals minute
      65 Lobes of lower leaves less than 3 cm,
      frequently and deeply pinnatifid
        8. Anthriscus sylvestris

      65 Lobes of lower leaves at least 4 cm, not
      or shallowly pinnatifid   HERACLEUM
        (a) Umbel with 10–20 rays
        67. H. sphondylium

        (a) Umbel with 50–150 rays
        68. H. mantegazzianum

    64 Plant glabrous; sepals conspicuous
      66 Bracts present   OENANTHE
        (a) Lobes of upper leaves lanceolate to
        ovate   **28. O. crocata**

(a) Lobes of upper leaves linear to spathulate

(b) Roots with tubers distant from base of stem; partial umbels flat-topped in fruit; rays and pedicels thickening in fruit   **25. O. pimpinelloides**

(b) Roots tuberous from their junction with the stem; partial umbels irregular in fruit; rays and pedicels not thickening in fruit   **27. O. lachenalii**

66 Bracts absent

67 Pedicels at least twice as long as fruit; stock septate   **53. Cicuta virosa**

67 Pedicels mostly much shorter than fruit; stock not septate   OENANTHE

(a) Lobes of upper leaves lanceolate to ovate   **30. O. aquatica**

(a) Lobes of upper leaves linear to spathulate

(b) Pinnate part of cauline leaves shorter than the hollow petiole   **24. O. fistulosa**

(b) Pinnate part of cauline leaves longer than the solid or flattened petiole

(c) Rays and pedicels thickening in fruit   **27. O. lachenalii**

(c) Rays and pedicels thickening in fruit   **26. O. silaifolia**

MULTI-ACCESS KEY

Table of alternative states

A   Flowers yellow or greenish-yellow
B   Flowers not yellow

C   Lowest leaves simple, entire or toothed
D   Lowest leaves lobed, 1-pinnate or 1-ternate
E   Lowest leaves at least 2-pinnate or 2-ternate

F   Fruit at least 3 times as long as wide
G   Fruit less than 3 times as long as wide

H   Fruit and ovary with hairs, spines, scales or papillae
I   Fruit and ovary without hairs, spines, scales or papillae

J   Fibres present and conspicuous around base of stem
K   Fibres absent

L   Umbel simple or flowers in whorls or a dense head
M   Umbel compound; bracteoles present
N   Umbel compound; bracteoles absent, rarely 1–2 and caducous

O   Annual
P   Biennial or perennial

Q   Stem with deflexed, closely appressed hairs
R   Stem glabrous or with hairs not deflexed and closely appressed

S   Fruit not winged
T   Fruit winged

Notes

It is sometimes difficult to distinguish between D and E, but doubtful species have been keyed out under both. S and T can present difficulties if the fruit is not fully ripe. In *Peucedanum* in particular the wings develop late, so if the fruit is unripe both alternatives should be tried.

Main Key

ACGIKMORS   BUPLEURUM (Key A)

ACGIKMPRS   BUPLEURUM (Key A)

ADGIKNPRT   Smelling of parsnip when crushed   **66. Pastinaca sativa**

AEGIJMPRT   **63. Peucedanum officinale**

AEGIKMPRS   Leaf-lobes subulate to linear-lanceolate   **22. Crithmum maritimum**

Lobes of lower leaves crenate or serrate, obtuse   **13. Smyrnium olusatrum**

Lobes of lower leaves pinnatifid, acute   **50. Petroselinum crispum**

AEGIKMPRT   Leaf-lobes 10–15 mm   **34. Silaum silaus**

Leaf-lobes usually 40–80 mm; bracteoles connate below   **62. Levisticum officinale**

AEGIKNORS   Strongly aromatic   **33. Anethum graveolens**

AEGIKNPRS   Leaf-lobes filiform   **32. Foeniculum officinale**

Leaf-lobes not filiform   SMYRNIUM (Key B)

BCGIKLPRS   HYDROCOTYLE (Key C)

BCGIKMORS   Sepals large   **12. Coriandrum sativum**

BDFHKLPRS   **3. Astrantia major**

BDFIKMPRS   **55. Falcaria vulgaris**

BDGHKLPRS   Leaves spiny   ERYNGIUM (Key D)

Leaves not spiny   **2. Sanicula europaea**

BDGHKMOQS   Outer petals strongly radiating   **Turgenia latifolia** (see **72**)

Outer petals not strongly radiating   TORILIS (Key E)

BDGHKMPQT   **69. Tordylium maximum**

BDGHKMPRT   HERACLEUM (Key J)

BDGIKMPRS   Bracts all 3-fid or pinnatisect   AMMI (Key F)

Lowest leaves with more than 20 pairs of lobes   **52. Carum verticillatum**

Aquatic, with translucent submerged

leaves; umbel with
1–2(–4) rays   **48. Apium inundatum**

Aquatic, with trans-
lucent submerged
leaves; umbel with
(4–)5–16 rays   OENANTHE (Key G)

Petioles fistular; partial
umbels subglobose
in fruit   **24. Oenanthe fistulosa**

Most umbels subsessile
and leaf-opposed   APIUM (Key H)

Longest bracts less than
$\frac{1}{2}$ as long as shortest
rays   **52. Sison amomum**

Stem hollow; petals
papillose beneath   **20. Sium latifolium**

Stem hollow; petals
smooth beneath   **21. Berula erecta**

Stem solid; petals
smooth beneath   **51. Petroselinum segetum**

BDGIKMPRT   Plant glabrous   **59. Ligusticum scoticum**

Peduncles hairy; leaves
glabrous   **65. Peucedanum
ostruthium**

Peduncles hairy; leaves
hairy   HERACLEUM (Key J)

BDGIKNPRS   Umbels mostly subsessile and leaf-
opposed   **45. Apium graveolens**

Leaves ternate; plant
with long rhizomes   **19. Aegopodium
podagraria**

Umbels long-pedunculate;
plant without
rhizomes   PIMPINELLA (Key K)

BEFIKMORS   **10. Scandix pecten-
veneris**

BEFIKMPRS   Strongly aromatic; stem
hollow   **11. Myrrhis odorata**

Midrib of leaf-lobes
(at least near the base)

| | | with a vein on either side close to and parallel with it | **55. Falcaria vulgaris** |
| | | Stems conspicuously swollen near the nodes | CHAEROPHYLLUM (Key L) |

with a vein on either
side close to and
parallel with it   **55. Falcaria vulgaris**

Stems conspicuously
swollen near the
nodes   CHAEROPHYLLUM (Key L)

BEGHJMPRS   **23. Seseli libanotis**

BEGHKMOQS   Sepals conspicuous, as
long as the petals   **Caucalis platycarpos
(see 72)**

Sepals shorter than
petals   TORILIS (Key E)

BEGHKMORS   **49. Trachyspermum ammi**

BEGHKMPRS   Central flower of umbel
usually dark purple   **73. Daucus carota**

BEGIJMPRS   Usually dioecious; S.
England, on lime-
stone   **44. Trinia glauca**

Not dioecious; N. Wales,
N. England, Scotland,
mountain grassland   **35. Meum athamanticum**

BEGIJNPRS   **44. Trinia glauca**

BEGIKMORS   **31. Aethusa cynapium**

BEGIKMPRS   Bracts all 3-fid or
pinnatisect   AMMI (Key F)

Stems usually purple-
spotted; bracteoles
on outer side of partial
umbels only   **37. Conium maculatum**

Leaves hairy on margin
and larger veins (lens);
fruit wider than
long   **36. Physospermum
cornubiense**

Leaves and stems hairy   ANTHRISCUS (Key M)

Stem flexuous and
tapering downwards
from soil level to its
junction with the tuber,
hollow after flowering;
styles suberect in
fruit   **16. Conopodium majus**

Stem flexuous and
tapering downwards
from soil level to its
junction with the tuber,
always solid; styles
recurved in fruit   **15. Bunium
bulbocastanum**

Pedicels at least twice
as long as long as
fruit; stock septate   **53. Cicuta virosa**

Most pedicels much
shorter than fruit   OENANTHE (Key N)

BEGIKMPRT   Stem solid   **58. Selinum carvifolia**

Umbels with fewer than
15 rays   **59. Ligusticum scoticum**

Leaf-lobes pinnatifid;
bracts present   **64. Peucedanum palustre**

Leaves ternate   **65. Peucedanum
ostruthium**

Stems usually pruinose   ANGELICA (Key O)

BEGIKNPRS   Umbels mostly with
fewer than 10 rays;
fruit smelling of
caraway when
crushed   **56. Carum carvi**

Rhizome far-creeping;
plant glabrous   **19. Aegopodium
podagraria**

Umbels long-pedunculate;
plant without
rhizomes   PIMPINELLA (Key K)

Subsidiary Keys

Key A.   *Bupleurum*

1 Perennial
  2 Herb; leaves parallel-veined   **42. B. falcatum**
  2 Shrub; leaves with a distinct midrib and network of lateral
    veins   **43. B. fruticosum**

1 Annual
   3 Upper cauline leaves perfoliate; bracts absent
      4 Fruit papillose   **39. B. subovatum**
      4 Fruit not papillose   **38. B. rotundifolium**
   3 Upper cauline leaves not perfoliate; bracts present
      5 Leaves parallel-veined, with distinct cross-veins; bracteoles subulate, not concealing the flowers   **41. B. tenuissimum**
      5 Leaves parallel-veined, without distinct cross-veins; bracteoles ovate, concealing the flowers   **40. B. baldense**

Key B. *Smyrnium*

Upper leaves not perfoliate   **13. S. olusatrum**
Upper leaves perfoliate   **14. S. perfoliatum**

Key C. *Hydrocotyle*

Leaves peltate, glabrous   **1. H. vulgare**
Leaves with a deep sinus, usually hairy   **H. moschata** (see **1**)

Key D. *Eryngium*

Leaves 3(–5)-lobed; flowers and often the stems bluish
                                   4. E. maritimum
Leaves pinnate; flowers whitish; stems pale green   **5. E. campestre**

Key E. *Torilis*

1 Umbels subsessile   **70. T. nodosa**
1 Umbels long-pedunculate
   2 Bracts 0–1; fruit 4–6 mm   **71. T. arvensis**
   2 Bracts 4–6(–12); fruit 3–3.5 mm   **72. T. japonica**

Key F. *Ammi*

Umbel with slender spreading rays after flowering   **54. A. majus**
Umbel with thickened contracted rays after flowering
                             A. visnaga (see **54**)

Key G.   *Oenanthe* (part)

Fruit 3.5–4.5 mm, ovoid   **30. O. aquatica**
Fruit 5–6.5 mm, cylindrical   **29. O. fluviatilis**

Key H.   *Apium* (part)

Stem rooting at lower nodes only; bracts 0–2   **46. A. nodiflorum**
Stem rooting at all nodes; bracts 3–7   **47. A repens**

Key J.   *Heracleum*

Umbel with 10–20 rays   **67. H. sphondylium**
Umbel with 50–150 rays   **68. H. mantegazzianum**

Key K.   *Pimpinella*

Stem hollow, prominently angled; lower leaves with 3–4 pairs of
  lobes   **17. P. major**
Stem solid, at least when young, terete; lower leaves with (2–)4–7
  pairs of lobes   **18. P. saxifraga**

Key L.   *Chaerophyllum*

Biennial; leaves dark green; fruit 5–6 mm   **6. C. temulentum**
Perennial; leaves yellowish; fruit c. 9 mm   **7. C. aureum**

Key M.   *Anthriscus*

1 Stem glabrous or very sparsely hairy; pedicels at least as thick as
  the rays in fruit   **9. A. caucalis**
1 Stem pubescent, at least above the nodes; pedicels thinner than
  the rays in fruit
    2 Rays glabrous   **8. A. sylvestris**
    2 Rays hairy   **A. cerefolium (see 8)**

Key N. *Oenanthe* (see also Key G)

1 Bracts present
  2 Lobes of upper leaves lanceolate to ovate   **28. O. crocata**
  2 Lobes of upper leaves linear to spathulate
    3 Roots with tubers distant from the base of the stem; partial umbels flat-topped in fruit; rays and pedicels thickening in fruit   **25. O. pimpinelloides**
    3 Roots tuberous from their junction with the stem; partial umbels irregular in fruit; rays and pedicels not thickening in fruit   **27. O. lachenalii**
1 Bracts absent
  4 Lobes of upper leaves lanceolate to ovate   **30. O. aquatica**
  4 Lobes of upper leaves linear to spathulate
    5 Pinnate part of cauline leaves shorter than the hollow petiole   **24. O. fistulosa**
    5 Pinnate part of cauline leaves longer than the solid or flattened petiole
      6 Rays and pedicels not thickening in fruit   **27. O. lachenalii**
      6 Rays and pedicels thickening in fruit   **26. O. silaifolia**

Key. O. *Angelica*

Flowers white or pinkish; stem usually purplish; fruit 4–5 mm   **60. A. sylvestris**
Flowers greenish; stem usually green; fruit c. 6 mm   **61. A. archangelica**

DESCRIPTIONS AND ILLUSTRATIONS

NOTES ON DESCRIPTIONS

Basal leaves are those at or near the base of the flowering stem, not leaves of seedlings.

The number and length of rays, bracts and bracteoles refers to main umbels with ripe fruit.

Fruit length excludes the stylopodium.

Styles and stigmas are described in ripe fruit.

The stylopodium is described for hermaphrodite flowers.

Descriptions of cotyledons refer to young seedling stage.

Abbreviations of authorities for plant names and of periodicals are those used in *Flora Europaea*.

A chromosome number followed by an asterisk indicates that the the count was made on material of British origin.

The main flowering period is given at the end of each description, but odd plants may flower earlier or later, the latter particularly when they have been cut down.

Distribution in the British Isles is given, usually in outline, and is derived from *Atlas of the British Flora* (ed. F. H. Perring and S. M. Walters, 2nd edition 1976); pre-1930 records, where these are distinguished, are omitted. Further details can be obtained by reference to the *Atlas*.

The size of each genus is indicated at the end of the account of the first species of each.

Introduced Species

Those introduced species which appear to be of fairly constant occurrence as casuals have been included. Many others have occurred on one or few occasions and changes in frequency are very liable to happen.

It should be borne in mind that casuals, especially on rubbish dumps, are often dwarfed and may not set fruit, so that identification may be difficult.

36

In each illustration the whole plant, or major parts of a plant (roots, leaves, inflorescences) when drawn separately, are shown at half natural size. Fruits are shown in greater detail, in both dorsal and lateral view where relevant, and a cross-section of one of the two carpels is also shown. The magnifications of the fruit, and also of flowers in a few cases, are indicated at the foot of the text page opposite.

1. Hydrocotyle vulgaris L.

Marsh Pennywort, White-rot

A perennial, glabrous except for the petioles, very variable in size, with slender creeping stems which root at the nodes. *Leaves* more or less orbicular in outline, 8–35 mm across, peltate, crenate, with 6–9 main veins radiating from the junction of blade and petiole; petioles up to 25 cm, sparsely hairy, with scarious often laciniate stipules. *Inflorescence* usually about half as long as the subtending leaf. *Bracteoles* present. *Flowers* greenish-white tinged with pink, hermaphrodite, subsessile, 3–6 in a simple, head-like umbel c. 3 mm across, sometimes with 1–3 whorls of flowers below; sepals minute or absent; ovary with a more or less flat disc at the apex; styles not thickened at the base. *Fruit* c. 2 mm wide, wider than long, covered with brownish resinous dots, strongly compressed laterally; commissure narrow; carpophore absent; mericarps with slender, rather prominent ridges; vittae absent, at least when the fruit is mature; styles spreading horizontally or somewhat recurved, about twice as long as the disc; stigma truncate. *Cotyledons* abruptly contracted into a petiole. $2n = 96$. Flowering in June and July.

Recorded from every vice-county in the British Isles, but absent from large areas of the Midlands and N. Scotland, probably owing to lack of suitable habitats. W., C. and S. Europe to c. 60°N in Scandinavia; Caucasus; records from N.W. Africa do not seem to have been confirmed recently.

It is reported to be used as a cure for whooping-cough in Danish folk medicine.

H. moschata G. Forster occurs locally as a weed in lawns and is well established on grassy banks on Valentia Island in S.W. Ireland. It is like *H. vulgaris* in general appearance but is aromatic and usually has leaves 10–20 mm across, hirsute on both surfaces, with a deep sinus and 5–7 shallow lobes which are distinctly toothed. The petioles are up to c. 5 cm, with dense, patent to deflexed hairs and there are usually 10–20 flowers in an umbel. Vegetatively *H. moschata* shows a considerable resemblance to *Sibthorpia europaea* but the shallow lobes of the leaves are distinctly toothed in the former and entire in the latter. It is native in New Zealand.

The genus contains about 100 species, widely distributed in temperate and tropical regions, especially in the Southern Hemisphere. The slender creeping stems and more or less orbicular leaves distinguish *Hydrocotyle* from all other British umbellifers.

Fruit × 15; fruit t.s. × 15

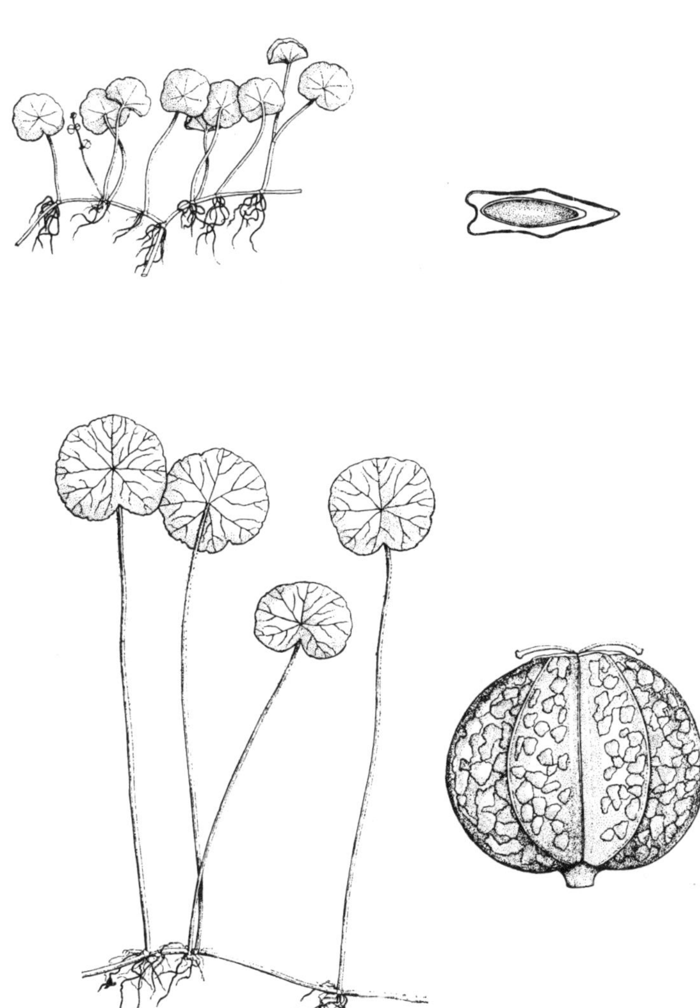

2. **Sanicula europaea** L.

Sanicle

A perennial with a stout stock. *Stems* 20–60 cm. *Leaves* 2–6 cm
across, deeply palmately lobed, usually 5-lobed, the lobes obovate-
cuneate, usually again shallowly 3-lobed and sharply serrate, with
the teeth ending in a bristle not more than 0.5 mm long. *Basal leaves*
long-petiolate, the cauline few or absent, shortly petiolate or sessile,
otherwise similar to the basal but smaller; stipules absent. *Inflorescence* of a number of simple umbels arranged in an irregular cyme
and often giving the appearance of a compound umbel. *Bracts*
simple, 3-fid or sometimes leaf-like; bracteoles shorter than the
flowers, simple or 3-fid. *Umbels* c. 5 mm across, subglobose, with
shortly pedicellate male flowers and 3–6 sessile hermaphrodite
flowers. *Flowers* white or pinkish; sepals conspicuous, longer than
the inflexed petals; ovary with a more or less flat disc at the apex;
styles not thickened at the base. *Fruit* c. 3 mm, subterete, covered
with rigid, forward-pointing, hooked bristles; commissure broad;
carpophore absent; vittae absent; styles c. 3 mm, divergent to somewhat recurved; stigma capitate. *Cotyledons* abruptly contracted into
a petiole. $2n = 16$. Flowering from May to August.

In woods, particularly of beech and oak. Widely distributed in
suitable habitats. Throughout most of Europe, Asia and Africa,
but rare and only on mountains in the south.

The genus contains about 40 species which are widely distributed
particularly in temperate regions and on mountains in the tropics,
except for Australia.

Fruit × 7; fruit t.s. × 7.

3. **Astrantia major** L.

Astrantia

A densely tufted glabrous perennial. *Stems* up to 100 cm, erect, usually branched, with the uppermost branches in a whorl of 3 or more. *Leaves* deeply palmately lobed, the basal long-petiolate, 6–17 cm across, 3- to 5(–7)-partite, with oblanceolate to obovate, irregularly toothed and often somewhat lobed segments, the middle segment free for at least $\frac{2}{3}$ its length. *Cauline* leaves smaller than the basal, usually few, shortly petiolate or the upper sessile. *Umbels* simple, the terminal with hermaphrodite and a few male flowers, the lateral umbels with mostly male flowers; pedicels of hermaphrodite flowers 2–5 mm, of male flowers 4–10 mm. *Bracteoles* as long as or longer than the flowers, erect, lanceolate, acute, pink or white, strongly reticulately veined. *Flowers* whitish or pinkish; sepals conspicuous, longer than the inflexed petals; ovary with a more or less flat disc at the apex; styles not thickened at the base. *Fruit* 6–8 mm, about 3 times as long as wide, oblong, covered with 2-dentate vesicular scales arranged in 5 longitudinal rows on each mericarp; commissure broad; carpophore absent; vittae solitary; styles 4–5 mm, divergent or somewhat recurved; stigma capitate. *Cotyledons* abruptly contracted into a petiole. $2n = 28$. Flowering from May to July.

Naturalized in scattered localities, chiefly in the north and west of Britain. Native of C. and E. Europe, extending to N. Spain and White Russia.

Represented in Britain by subsp. **major.**

The genus contains about 10 species, which occur in Europe and S.W. Asia.

Fruit × 5; fruit t.s. × 5.

4. **Eryngium maritimum** L.

Sea Holly

An intensely glaucous perennial. *Stems* 15–60 cm, branched and usually bluish above, rigid, solid. *Basal leaves* 4–10 cm, coriaceous, suborbicular, truncate or cordate at the base, 3(–5)– lobed, with thick prominent veins, a thick cartilaginous margin and large spinose teeth; petiole as long as the lamina. *Cauline leaves* similar, but smaller and sessile. *Inflorescence* of pedunculate, subglobose capitula 1.5–3 cm. *Bracts* about as long as the capitulum, ovate or ovate-lanceolate, similar to the leaves in texture, spininess and colour; bracteoles tricuspidate, spiny, usually purplish-blue, longer than the flowers. *Flowers* bluish-white; sepals 4–5 mm, longer than the petals, lanceolate, with a prominent midrib, which is excurrent as a stout spine; ovary with a more or less flat disc at the apex; styles slightly thickened at the base. *Fruit* 13–15 mm, scarcely compressed, covered in papillae which become longer towards the apex; commissure broad; carpophore absent; vittae very slender, styles c. 6 mm, divergent to somewhat recurved; stigma tapering. *Cotyledons* abruptly contracted into a petiole. $2n = 16$. Flowering from June to September.

On sand-dunes or less frequently shingle a little above high-tide mark. Around the coasts of the British Isles but apparently extinct in N.E. England north of Flamborough Head and in E. Scotland. Coasts of Europe north to c. 60°N, and of North Africa and S.W. Asia.

The roots were formerly candied as a sweetmeat and are recommended by Dioscorides as a remedy for flatulence. The young shoots have been eaten like asparagus.

The genus is almost cosmopolitan, and contains about 230 species.

Fruit × 5; fruit t.s. × 4.

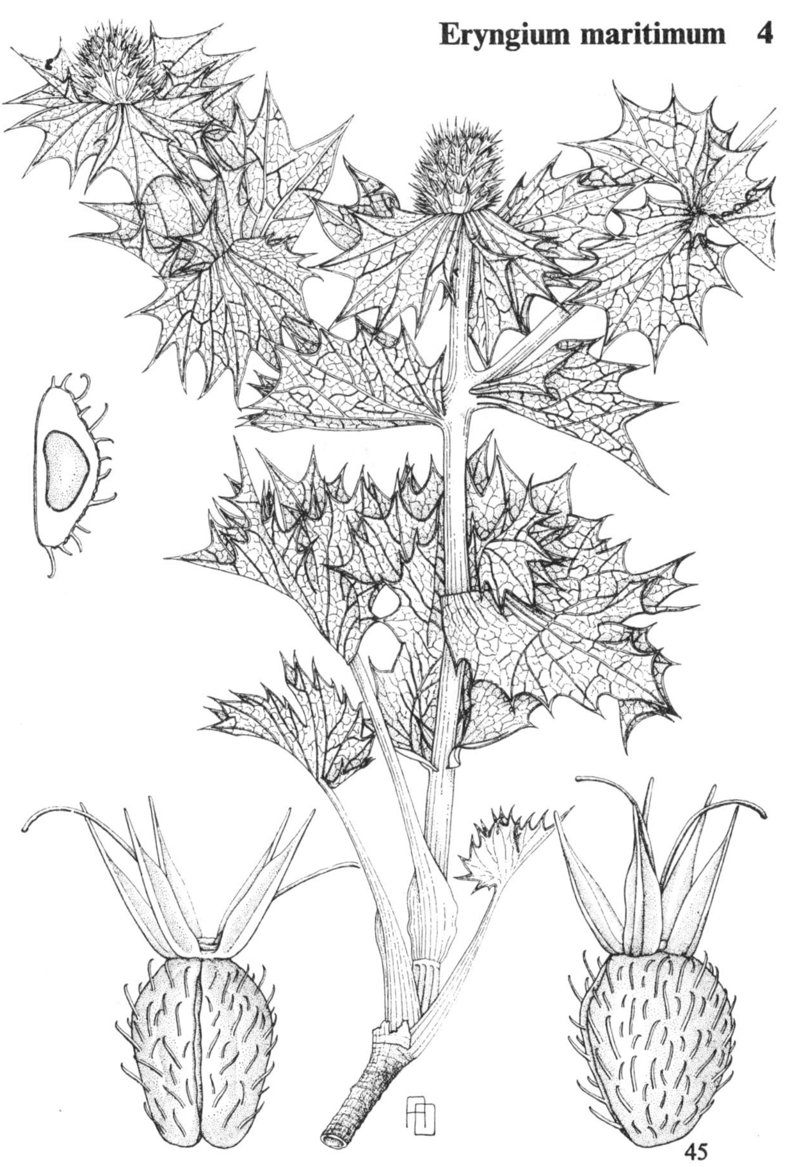

45

5. **Eryngium campestre** L.

Field Eryngo

A much-branched pale green, glaucous perennial. *Stems* 30–60 cm, rigid, solid. *Basal leaves* 5–20 cm, coriaceous, triangular-ovate in outline, pinnately divided, the primary divisions decurrent on the rhachis; lobes spinose-serrate, with thick prominent midrib ending in a stout spine, and a thick cartilaginous margin; petiole as long as the lamina. *Cauline leaves* progressively smaller and less divided, sessile, with a broad, spiny-margined semi-amplexicaul base. Flowers in usually very numerous pedunculate, ovoid capitula mostly 1–1.5 cm. *Bracts* 5–8, 1½–3 times as long as the capitulum, linear-lanceolate, with spinose apex, entire or with 1(–2) pairs of lateral spines; bracteoles entire, longer than the flowers. *Flowers* white; sepals c. 2.5 mm, longer than the petals, lanceolate, with a prominent midrib which is excurrent as a stout spine; ovary with a more or less flat disc at the apex; styles slightly thickened at the base. *Fruit* c. 5 mm, scarcely compressed, densely covered in very acute white scales; commissure broad; carpophore absent; vittae very slender; styles c. 4 mm, divergent to somewhat recurved; stigma tapering. *Cotyledons* abruptly contracted into a petiole. $2n = 14, 28$. Flowering in July and August.

Open habitats, mostly near the sea. Native or long naturalized in a few localities in southern England and Wales; rare. S. and C. Europe, North Africa, S.W. Asia.

Fruit × 10; fruit t.s. × 5.

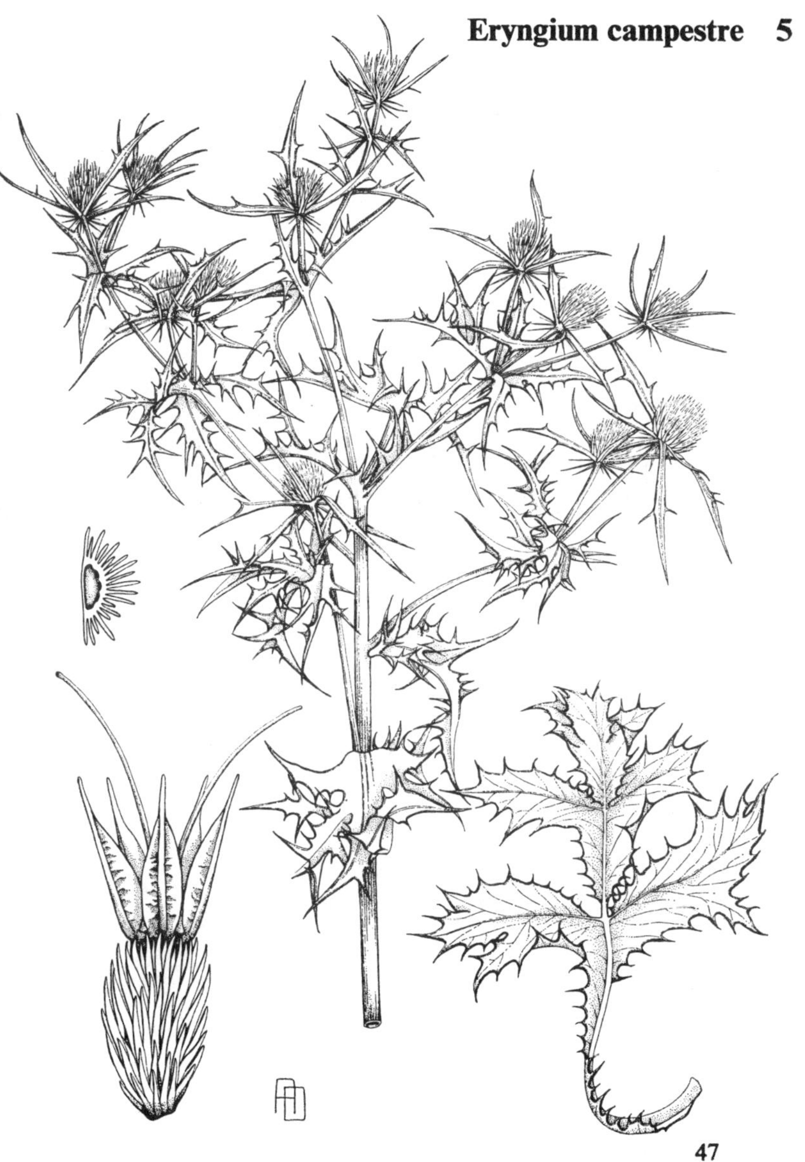

6. Chaerophyllum temulentum L.

Rough Chervil

A somewhat hispid biennial. *Stems* up to 100 cm, solid, swollen below the nodes, purple-spotted or entirely purple. *Leaves* 2- to 3-pinnate, dark green, appressed-hairy on both surfaces, long-petiolate: lobes mostly 10–20 mm, ovate in outline, deeply toothed, the teeth abruptly contracted at the apex. *Umbels* compound, with (4–)6–12(–15) hairy rays usually 1.5–5 cm long; peduncle longer than the rays, hairy; terminal umbel with mostly hermaphrodite flowers, overtopped by the lateral umbels, which have mostly male flowers. *Bracts* absent, or rarely 1 or 2; bracteoles 5–8, shorter than the pedicels, ciliate, eventually deflexed. *Flowers* white; sepals absent; outer petals not radiating; styles with enlarged base, forming the stylopodium. *Fruit* usually 5–6 mm, slightly compressed laterally, oblong but narrowed towards the apex, constricted at the commissure; mericarps with broad, rounded ridges; carpophore present; vittae solitary, conspicuous; pedicels without a ring of hairs at the apex; styles about as long as the stylopodium, recurved; stigma capitate. *Cotyledons* tapered gradually at the base, without a distinct petiole. $2n = 14, 22$. Flowering from late May to early July.

In hedges and other grassy places. Common in most of England and Wales, local and mainly in the east of Scotland and Ireland. Most of Europe, but rare in the Mediterranean region; S.W. Asia; N.W. Africa.

In much of England this is the commonest roadside umbellifer to flower just after *Anthriscus sylvestris* and is easily recognized by the purple-spotted stems and swollen top of the internodes. Poisonous.

The genus contains about 40 species, widely distributed in north temperate regions; it is thus one of the larger genera in subfamily Apioideae.

Fruit × 8; fruit t.s. × 10.

7. **Chaerophyllum aureum** L.

Golden Chervil

A robust, more or less hairy perennial. *Stems* up to 150 cm, solid, swollen below the nodes, sometimes purple-spotted. *Leaves* 3-pinnate, pale yellowish-green, appressed-hairy to nearly glabrous on both surfaces, long-petiolate; lobes usually 10–40 mm, lanceolate in outline, deeply toothed or lobed, the teeth gradually narrowed at the apex. *Umbels* compound, with 12–25 nearly or quite glabrous rays usually 1.5–3(–4.5) cm long; peduncle usually longer than the rays, hairy; terminal umbel with mostly hermaphrodite flowers, over-topped by the lateral umbels, which have mostly male flowers. *Bracts* absent or rarely 1–3; bracteoles 5–8, about as long as the pedicels in flower, hairy, ultimately deflexed. *Flowers* pure white; sepals minute; outer petals not radiating; styles with enlarged base, forming the stylopodium. *Fruit* usually 9 mm, oblong but narrowed rather abruptly near the apex, constricted at the commissure; mericarps with broad, rounded ridges; carpophore present; vittae solitary; pedicels without a ring of hairs at the apex; styles about twice as long as the stylopodium, recurved; stigma capitate. *Cotyledons* tapered gradually at the base, without a distinct petiole. $2n = 22$. Flowering in June and July.

Naturalized in a few places in S. Scotland and S. England. Native in C. and S. Europe and S.W. Asia.

Fruit × 5; fruit t.s. × 5.

51

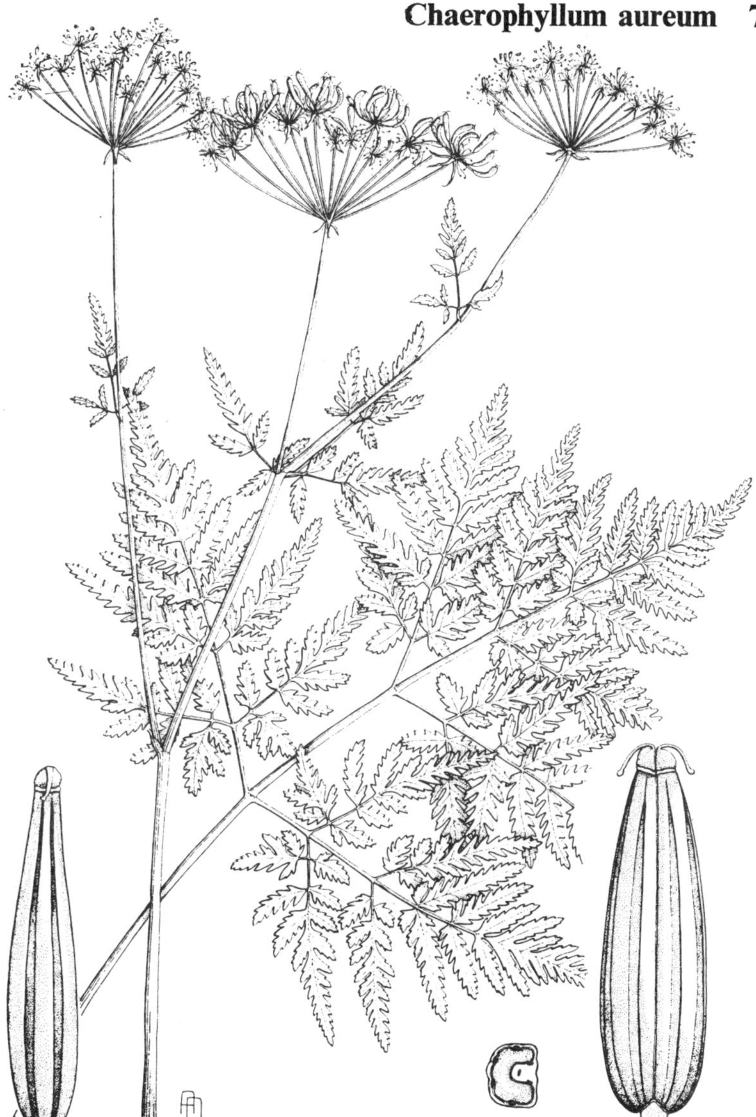

8. **Anthriscus sylvestris** (L.) Hoffm.

Cow Parsley, Keck

Plant more or less hairy, perennating by buds in the axils of the basal leaves, which develop a taproot and separate from the flowering stem when it dies. *Stems* 60–150 cm, hollow, furrowed, more or less puberulent. *Leaves* 3-pinnate, the lowest primary divisions much smaller than the rest of the leaf: lobes usually 10–30 mm, pinnatifid and often serrate. *Umbels* compound, with (3–)6–12 glabrous rays usually 1.5–3 cm long; peduncle about as long as the rays, more or less glabrous; terminal umbel with male flowers in the middle of each partial umbel surrounded by hermaphrodite flowers, the lateral umbels with mostly or entirely male flowers, equalling or slightly overtopping the terminal umbel. *Bracts* absent; bracteoles 4–6, ovate, aristate, ciliate, patent or deflexed. *Flowers* creamy white; sepals minute; outer petals not radiating; styles with enlarged base forming the stylopodium. *Fruit* 6–9 mm, oblong-ovoid, constricted at the commissure; mericarps smooth, with a very short, ridged beak; carpophore present; vittae solitary; pedicels with a ring of short, stout hairs at the apex; styles a little longer than the stylopodium, divergent; stigma capitate. *Cotyledons* tapered gradually at the base, without a distinct petiole. $2n = 16^*$. Flowering from April to early June.

Hedgerows, wood margins and waste places throughout most of the British Isles. Europe (rare in the south), temperate Asia and North Africa.

The commonest of the early-flowering umbellifers in much of England. In southern England the leaves have appressed hairs on the upper surface, giving them a rather pale green matt appearance, but in the north they are usually glabrous or nearly so and consequently appear darker green and rather shiny.

A. sylvestris can be easily distinguished from *Myrrhis odorata*, which grows in similar places in the north and flowers at the same time, by the smell when crushed and by the sharply angled fruits of the latter.

A. sylvestris is very resistant to herbicides.

The genus contains about 20 species in Europe, North Africa and temperate Asia.

Fruit $\times$ 5; fruit t.s. $\times$ 10.

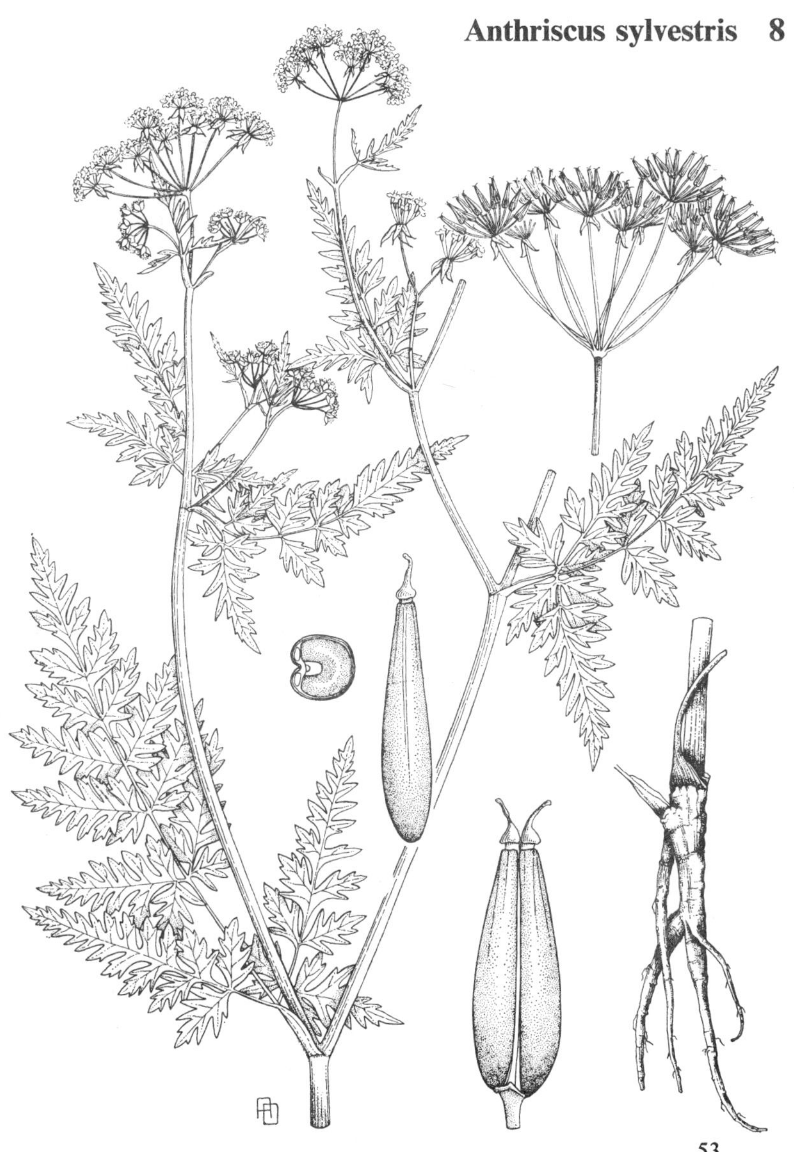

9. **Anthriscus caucalis** Bieb.

Bur Chervil

Annual; stems up to 80(–100) cm, usually freely branched, hollow, glabrous to very sparsely hairy. *Leaves* 2- to 3-pinnate, the lobes mostly 5–10 mm, dentate or pinnatified, hairy beneath, usually glabrous above; petioles inflated and villous towards the base. *Umbels* compound, with 3–6 glabrous rays usually c. 1 cm long; peduncle shorter than the rays, glabrous; umbels all leaf-opposed and with all flowers hermaphrodite. *Bracts* absent, rarely 1; bracteoles usually 4–5, ovate, acuminate, ciliate. *Flowers* white; sepals absent; outer petals not radiating; styles with enlarged base, forming the stylopodium. *Fruit* c. 3 mm, ovoid, shortly beaked, covered in hooked spines; mericarps with ridges only in the beak; commissure narrow; carpophore present; vittae absent; pedicels becoming at least as thick as the rays in fruit, with a ring of stout hairs at the apex; styles very short, connivent; stigma slightly thickened. *Cotyledons* tapered gradually at the base, without a distinct petiole. $2n = 14$. Flowering in May and June.

In hedgebanks, sandy ground near the sea and in waste places, rather local and absent from much of the north and west. W., C. and S. Europe, temperate Asia and North Africa; mainly on mountains in the south.

Var. **caucalis**, described above, is the only variant which occurs in this country. Var. *neglecta* (Boiss. & Reuter) P. Silva & Franco has smooth fruit.

A. cerefolium (L.) Hoffm., Garden Chervil, was formerly naturalized near Ross-on-Wye. It is rather like *A. caucalis* but has hairy rays and fruits 7–10 mm long, with no hooked spines. It is sometimes cultivated for flavouring and may persist for a short time.

Fruit × 8; fruit t.s. × 12.

10. Scandix pecten-veneris   L.

Shepherd's-Needle

A sparsely hairy annual. *Stems* up to 50 cm, becoming hollow when old. *Leaves* 2- to 3-pinnate, the lobes up to c. 10 mm, narrow, entire to pinnatifid; petiole widened at the base, with a scarious, usually ciliate margin. *Umbels* with 1–3 stout, glabrous or sparsely hairy rays 0.5–4 cm long; peduncle very short or absent; terminal umbel with hermaphrodite flowers, the lateral umbels with varying proportions of male and hermaphrodite flowers. *Bracts* usually absent; bracteoles usually 5, longer than the pedicels, simple or irregularly and often deeply divided. *Flowers* white; sepals small; outer petals not radiating; styles with enlarged base, forming the stylopodium. *Fruit* 30–70 mm, more or less cylindrical, slightly compressed laterally, with a strongly dorsally flattened beak 3–4 times as long as and clearly distinct from the seed-bearing portion, constricted at the commissure; mericarps ribbed, scabrid and with forward-pointing bristles on the margins; carpophore present; vittae solitary, conspicuous; pedicels almost as thick as the rays, glabrous at the apex; styles 2–4 times as long as the stylopodium, erect; stigma tapering. *Cotyledons* tapered gradually at the base, without a distinct petiole. $2n = 16, 26$. Flowering in May and June.

Formerly widely distributed as a weed of arable land in the south-eastern half of England but now rather rare; very local and mainly near the coast elsewhere. W., C. and S. Europe, W. and C. Asia, North Africa.

The above description refers to subsp. **pecten-veneris,** the only subspecies in Britain.

S. australis L. occurs as a casual, but does not persist. It may be distinguished from *S. pecten-veneris* by the slightly laterally compressed beak which is not clearly distinct from the seed-bearing portion.

The genus contains about 15 species, mostly confined to the Mediterranean region.

Fruit × 1; fruit t.s. × 7.

11. **Myrrhis odorata** (L.) Scop.

Sweet Cicely

A softly and rather sparingly hairy perennial, smelling of aniseed when crushed. *Stems* 60–200 cm, hollow. *Leaves* 2- to 3-pinnate, pale beneath and usually with some whitish markings; lobes usually 10–30 mm, oblong-lanceolate, pinnatifid or deeply serrate; cauline leaves with narrow sheathing petioles. *Umbels* compound, with 4–10(–21) hairy rays usually 1.5–3 cm, those bearing male flowers slender, the others stout; peduncles usually longer than the rays, hairy; terminal umbel with male and hermaphrodite flowers, the lateral usually with male flowers only. *Bracts* usually absent; bracteoles c. 5, lanceolate, with a long slender apex, whitish. *Flowers* white; sepals minute; outer petals slightly radiating; styles with enlarged base, forming the stylopodium. *Fruit* 15–25 mm, linear-oblong, with a short beak, sharply angled, with short forward-pointing bristles on the angles, dark brown and shiny when mature; commissure broad; carpophore present; vittae slender or absent; pedicels c. 5 mm, sparsely hairy; styles much longer than the stylopodium, patent or recurved in fruit; stigma capitate. *Cotyledons* tapered gradually at the base, without a distinct petiole. Flowering in May and June.

Hedge-banks and other grassy places especially near houses. Naturalized and locally abundant in N. England, much of Scotland and N. Ireland; very local or absent elsewhere. Native of the Alps, Pyrenees, Apennines and mountains of the western part of the Balkan peninsula; widely naturalized elsewhere.

Aromatic and stimulant; formerly used as a pot-herb. The smell when crushed and the whitish-flecked leaves distinguish it from all other British umbellifers.

M. odorata is the only species of the genus.

Fruit × 1.5; fruit t.s. × 4.

Myrrhis odorata 11
59

12. Coriandrum sativum L.

Coriander

A glabrous fetid annual. *Stems* up to 50(–70) cm, solid, almost smooth. *Lower leaves* long-petiolate, orbicular, 3-fid, or simply pinnate, with broad incise-serrate lobes c. 10 mm long, soon withering. *Cauline leaves* 2- to 3-pinnate, with narrow lobes. *Umbels* compound, with 2–5(–10) rays, usually 0.5–1.5 cm; peduncle usually longer than the rays, sometimes leaf-opposed; all umbels with a mixture of male and hermaphrodite flowers. *Bracts* usually absent; bracteoles usually 3, linear. *Flowers* purplish or white; sepals conspicuous, unequal; outer petals radiating; styles with enlarged base, forming the stylopodium. *Fruit* 2–6 × 2–5.5 mm, ovoid or globose, hard; mericarps not separating at maturity, with primary and secondary ridges, both low, the former narrower than the latter; commissure broad; carpophore present; vittae under the secondary ridges, obscure; pedicels of hermaphrodite flowers 4–5 mm, of male flowers shorter; styles much longer than the stylopodium, recurved in fruit; stigma capitate. *Cotyledons* tapered gradually at the base, without a distinct petiole. $2n = 22$. Flowering from July to October.

Cultivated for its aromatic fruits; a rare casual on waste ground. Perhaps native in North Africa and W. Asia.

Young plants can be recognized by the basal leaves and the unpleasant smell when they are crushed. Older plants show the ovoid to globose fruit with conspicuous sepals.

The genus contains this and one other species, *C. tordylium* (Fenzl) Bornm., in which the mericarps eventually separate, occurring in W. Asia.

Fruit × 5; fruit t.s. × 4.

60

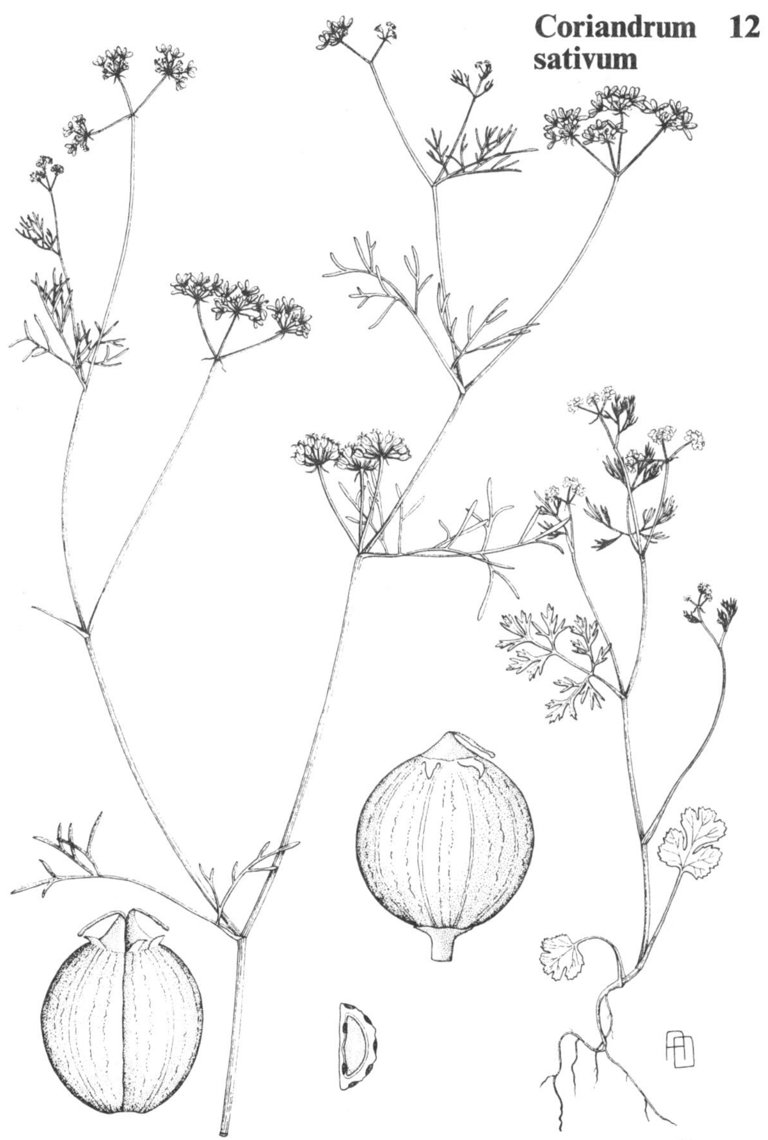

13. **Smyrnium olusatrum** L.

Alexanders

A glabrous biennial with a tuberous tap-root in the first year. *Stems* 50–150 cm, stout, solid, becoming hollow, distinctly ridged, especially above; branches often opposite above. *Leaves* 2- to 3-pinnate or -ternate, dark green and shiny, the lobes usually 25–80 mm, ovate to rhombic, serrate and sometimes lobed, the teeth obtuse; upper cauline leaves often opposite, ternate, with a greatly expanded sheathing petiole. *Umbels* compound, with usually 4–15 glabrous rays 3–4 cm long; peduncle usually longer than the rays, glabrous; terminal umbel with male and hermaphrodite flowers, the lateral umbels with mostly male flowers. *Bracts* and bracteoles few, small, or sometimes absent. *Flowers* yellow; sepals very small, somewhat accrescent in fruit; outer petals not radiating; styles with enlarged base, forming the stylopodium. *Fruit* 7–8 mm, broadly ovoid, laterally compressed, black, constricted at the commissure; mericarps with 3 prominent sharp ridges; carpophore present; vittae numerous; pedicels somewhat longer than the fruit, papillose on the inner angles; styles little longer than the stylopodium, at first patent, usually recurved and appressed to the stylopodium in ripe fruit; stigma capitate. *Cotyledons* abruptly contracted into a petiole. $2n = 22^*$. Flowering from April to June.

On hedge-banks, cliffs, and grassy roadsides. Well naturalized especially near the sea. Native in Europe, northwards to N.W. France; W. Asia, North Africa, Macaronesia.

Used as a pot-herb in some countries and formerly in this country. The young stems have a taste similar to celery.

The yellow flowers, the large, dark green shiny leaves with large lobes and the large black strongly and sharply ridged fruit serve to distinguish this from other British umbellifers.

The genus contains about 8 species, occurring in Europe and the Mediterranean region.

Fruit × 5; fruit t.s. × 4.

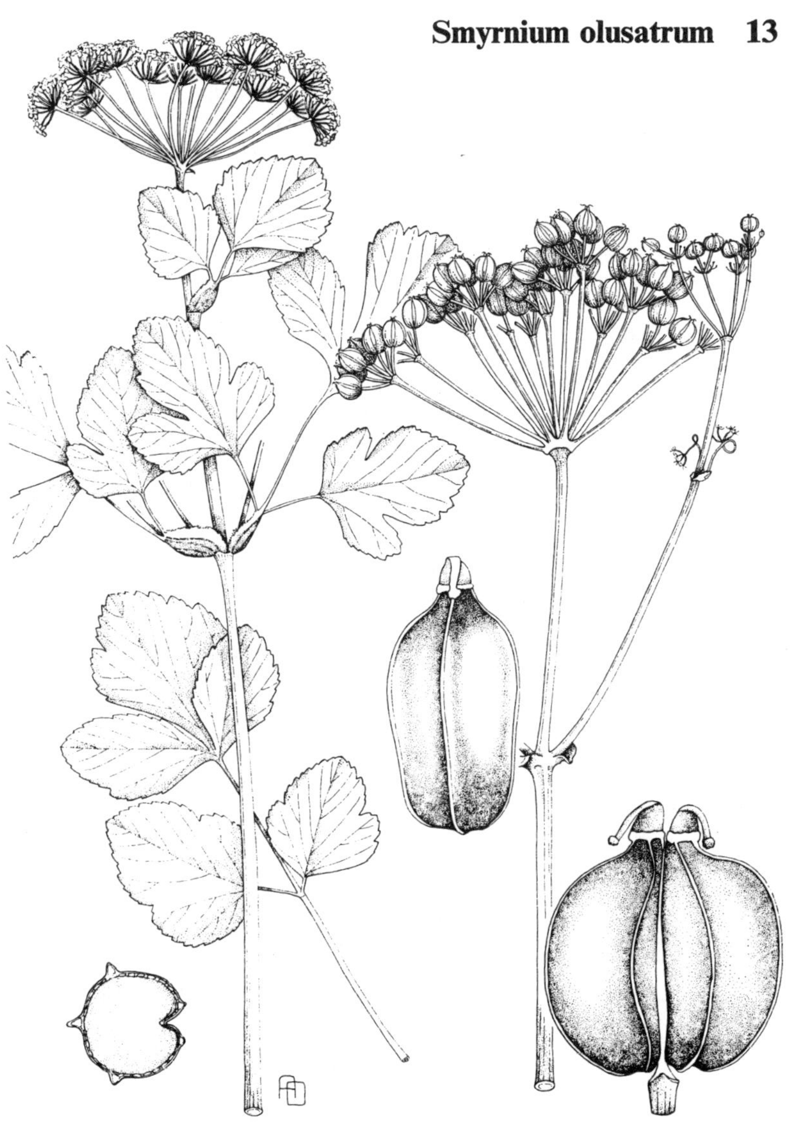

14. *Smyrnium perfoliatum* L.

An almost glabrous biennial with a tuberous tap-root. *Stems* (20–) 40–150 cm, solid, strongly angled and narrowly winged on the angles, particularly near the middle, the wings with rather sparse, small stellate hairs. *Basal leaves* 2- to 3-pinnate or -ternate, the lobes ovate, dentate or somewhat lobed. *Cauline leaves* 3–10 cm, simple, ovate, amplexicaul, crenate-serrate, rarely almost entire, alternate. *Umbels* compound, with 5–10 glabrous rays, which are papillose near the apex and 1–4 cm long; peduncle longer or shorter than the rays, winged on the angles; terminal umbel with mostly hermaphrodite flowers and the lateral umbels with mostly male flowers. *Bracts* and bracteoles absent. *Flowers* yellow; sepals very small, scarcely accrescent in fruit; outer petals not radiating; styles with enlarged base, forming the stylopodium. *Fruit* 3–3.5 × 5–5.5 mm, brownish-black, laterally compressed, constricted at the commissure; mericarps with 3 slender dorsal ridges; carpophore present; vittae numerous, slender; styles longer than the stylopodium, recurved and appressed in fruit; stigma capitate. *Cotyledons* abruptly contracted into a petiole. $2n = 22$. Flowering in May.

Sometimes cultivated for ornament and persisting as a weed in flower beds in South Shields, Chelsea and perhaps elsewhere. Established in long grass in the Queen's Cottage grounds, Kew, and in the Cambridge University Botanic Garden. Native of S. Europe northwards to E. Czechoslovakia; S.W. Asia, North Africa.

The root resembles a small turnip and the seedling has a tuberous hypocotyl. The crenate-serrate, amplexicaul cauline leaves distinguish this from all other British umbellifers.

Fruit × 7; fruit t.s. × 4.

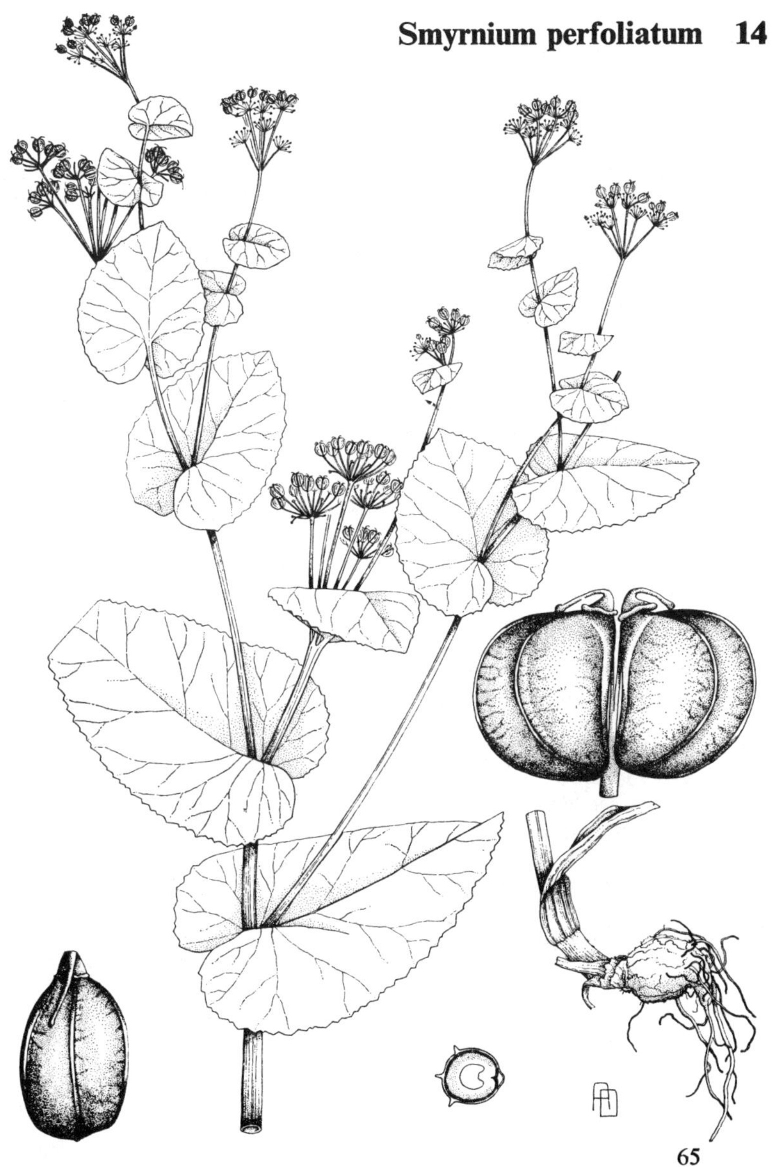

15. Bunium bulbocastanum L.

Great Pignut

A glabrous perennial. *Stems* 30–100 cm, usually erect, solid, arising from a solitary, globose, dark brown tuber 1–2.5 cm in diameter, the subterranean part of the stem flexuous and tapering downwards. *Leaves* mostly arising from the lower part of the stem, broadly triangular in outline, 3-pinnate, mostly withered at flowering time; lobes usually 5–10 mm, linear, with a cartilaginous apex; uppermost leaves 2-pinnate; petiole of lower leaves long, slender and abruptly expanded into a sheathing base; petiole of upper leaves wide and sheathing throughout. *Umbels* compound, with (5–)10–20 rays 1.5–4.5 cm long, scabrid on the angles; peduncle longer than the rays; most umbels with hermaphrodite and a few male flowers, the latest developed laterals with all male flowers. *Bracts* 5–10, linear-lanceolate, acuminate; bracteoles 5–10, linear to lanceolate. *Flowers* white; sepals a minute rim; outer petals not radiating; styles with enlarged base, forming the stylopodium. *Fruit* 3.5–4.5 mm, oblong-ellipsoid, laterally compressed, constricted at the commissure; mericarps with slender, pale ridges and solitary conspicuous reddish-brown vittae; carpophore present; pedicels 2–6 mm; styles about as long as the stylopodium, recurved in fruit; stigma capitate. *Cotyledon* tapered gradually at the base, without a distinct petiole. $2n = 22$. Flowering in June and July.

Very local in chalk grassland in Hertfordshire, Buckinghamshire, Bedfordshire and Cambridgeshire. Europe, northwards to England and eastwards to C. Germany and N.W. Jugoslavia; N.W. Africa.

The seedling has only one cotyledon, through abortion of the other during development. The tuber arises from the hypocotyl of the seedling and is edible raw or cooked. In the interests of conservation, please refrain from eating it.

Bunium is like *Conopodium* but the stem is solid after flowering and the styles are recurved in fruit. The stylopodium is sharply marked off from the styles, while in *Conopodium* it tapers gradually into them.

The genus contains about 40 species and extends from Europe and North Africa to C. Asia.

Fruit × 10; fruit t.s. × 10.

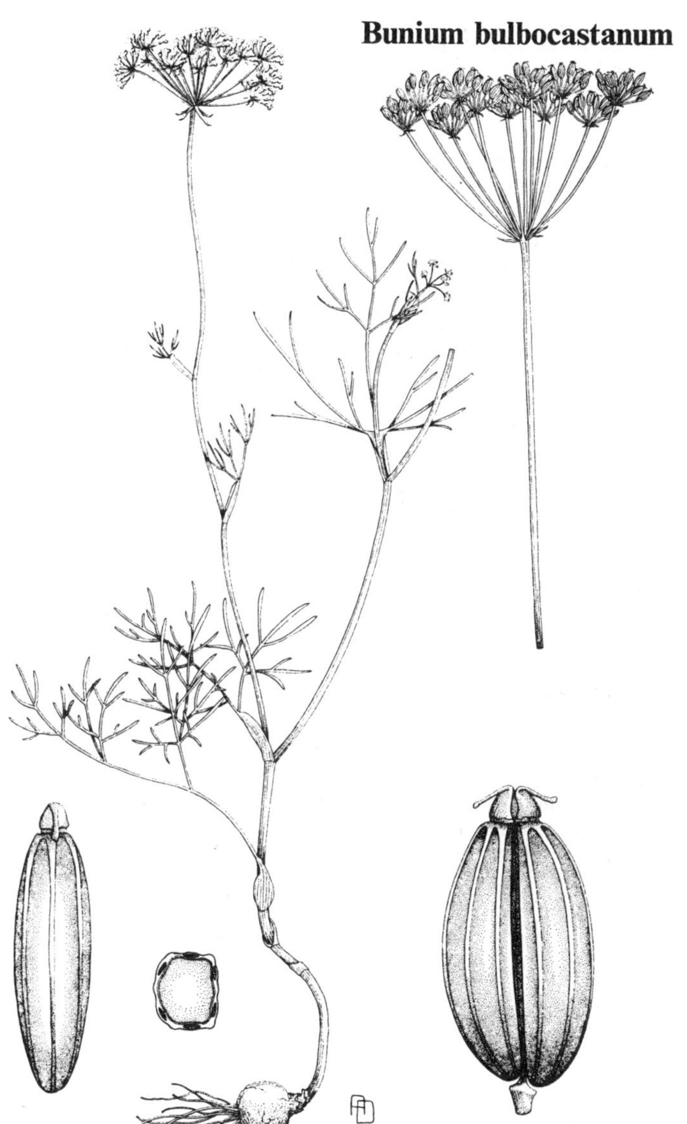

16. **Conopodium majus** (Gouan) Loret

Pignut

A glabrous perennial. *Stems* 8–60(–90) cm, hollow after flowering, arising from a solitary, usually rather irregular, dark brown tuber 1–3.5 cm across, the subterranean part of the stem flexuous and tapering downwards. *Leaves* mostly basal, long-petiolate, soon withering, the cauline few, the lower part of the stem leafless; basal leaves 3-pinnate, triangular in outline, the lobes usually 3–5 mm, linear-lanceolate, acute; cauline leaves 2-pinnate, with linear lobes, the terminal much longer than the lateral; petiole short, sheathing. *Umbels* compound, with 6–12 smooth rays 1–6 cm long; peduncle longer than the rays; umbels with mostly hermaphrodite and few male flowers, the latest developed laterals with male flowers. *Bracts* absent, rarely 1 or 2; bracteoles 2–5, linear. *Flowers* white; sepals absent; outer petals not or scarcely radiating; styles with enlarged base, forming the stylopodium. *Fruit* c. 4 mm, ovoid-oblong, almost terete, constricted at the commissure; mericarps with slender ridges and 2–3 reddish-brown vittae in each groove; carpophore present; pedicels longer than the fruit; styles about as long as the stylopodium, erect or divergent in fruit; stigma truncate. *Cotyledons* tapered gradually at the base, without a distinct petiole. Flowering in May and June.

In grassy places, hedge-banks, woods etc. Widely distributed and often common in Great Britain, less frequent in Ireland. Absent from the fens and scarce on chalk soils. Confined to W. Europe from Norway southwards, extending eastwards to Italy.

Unlike *Bunium*, both cotyledons develop. The tuber is edible, raw or cooked, and has a pleasant nutty flavour.

There are about 20 species in the genus, which occurs in Asia and North Africa, as well as Europe.

Fruit × 7; fruit t.s. × 10.

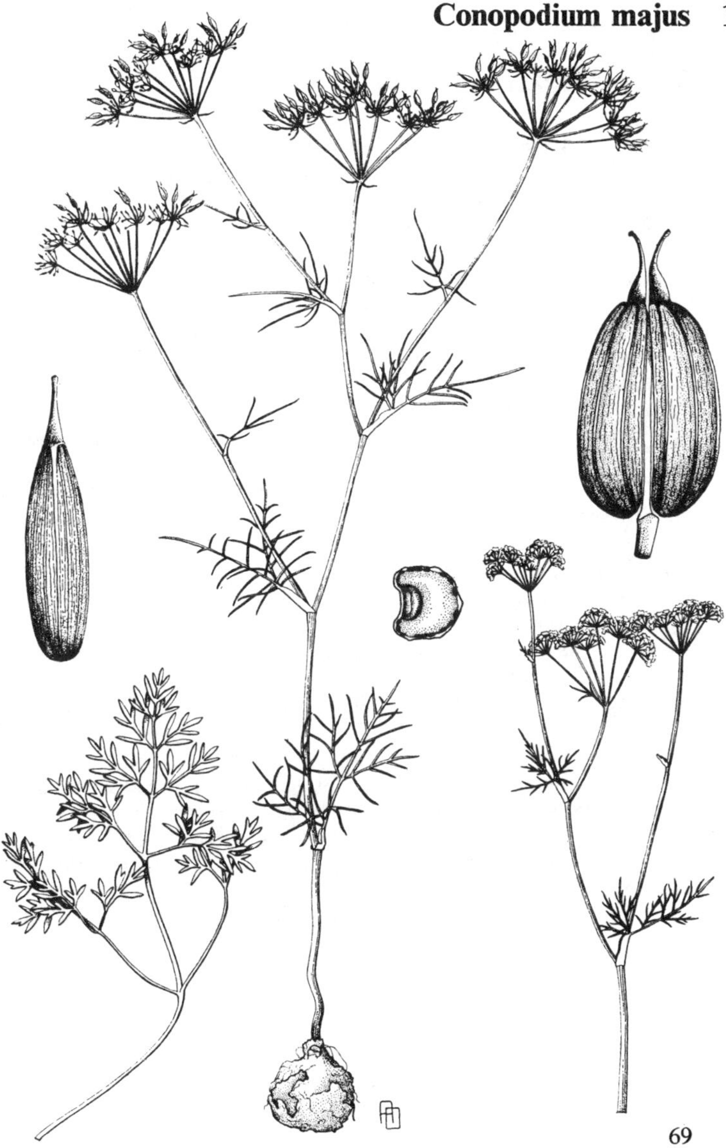

17. **Pimpinella major** (L.) Hudson

Greater Burnet-saxifrage

A nearly glabrous perennial. *Stems* up to 120 cm, arising from a stout stock with few or no fibrous remains of petioles, rather stout, prominently ridged or angled, often purplish towards the base, hollow. *Leaves* simply pinnate (very rarely 2-pinnate), long-petiolate, usually strigulose on the veins, at least beneath, and sparsely hirsute between them; lobes of lower leaves 3–4 pairs, up to 60(–100) mm, shortly stalked, ovate, usually truncate or subcordate at base, coarsely serrate or sometimes lobed, the teeth usually with fine cartilaginous points; upper cauline leaves with 1–2(–3) pairs of lobes, similar to the lower but smaller and often cuneate at base; petiole short and sheathing. *Umbels* compound, with 10–20 smooth rays 1–4 cm long; peduncle longer than the rays; terminal umbel with mostly hermaphrodite flowers, the lateral with mostly or entirely male flowers. *Bracts* absent; bracteoles usually absent. *Flowers* white or pinkish; sepals absent; outer petals not radiating; styles with enlarged base, forming the stylopodium. *Fruit* 3–4 mm, ovoid, laterally compressed; commissure broad; mericarps with prominent, slender ridges paler than the usually 3 reddish-brown vittae; carpophore present; pedicels c. 5 mm, glabrous; styles nearly as long as the petals and erect in flower, elongating later but often breaking off as the fruit ripens; stigma capitate. *Cotyledons* abruptly contracted into a petiole. $2n = 20$. Flowering from mid-July to early August.

Locally common on grassy roadsides and at the margins of woods from Northumberland southwards, but mainly in the eastern half of England. Most of Europe, except the extreme north and south.

The genus is a large one for subfam. Apioideae, containing about 150 species, mainly in north temperate regions but extending into the tropics.

Fruit × 12; fruit t.s. × 10.

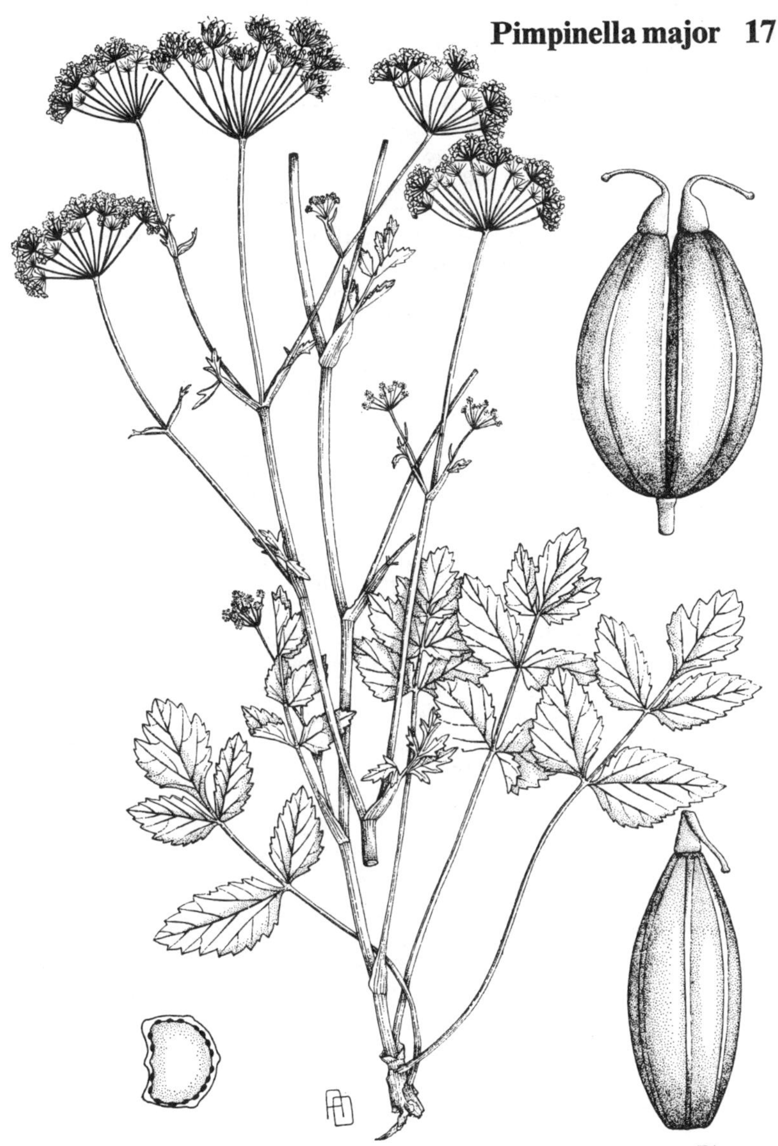

71

18. Pimpinella saxifraga   L.

Burnet-Saxifrage

Perennial, usually minutely hairy but sometimes glabrous. *Stems* (15–)30–100 cm, arising from a stock which often has fibrous remains of petioles, rather slender, terete, solid, developing a small hollow in fruit. *Leaves* very variable, the lowest and the upper usually simply pinnate, the intermediate usually 2-pinnate, more or less densely hairy or glabrous; lobes of lower leaves (2–)4–6(–7) pairs, up to c. 25 mm, broadly ovate to lanceolate in outline, coarsely serrate to pinnatifid, the teeth with very slender cartilaginous points; upper cauline leaves with up to 3 pairs of small, very narrow lobes; petiole longer than the lobes, inflated, usually purplish. *Umbels* compound, with 10–22 smooth rays (1–)2–3(–4.5) cm long; peduncle longer than the rays; umbels mostly with hermaphrodite flowers. *Bracts* and bracteoles absent. *Flowers* white, rarely pinkish; sepals absent; petals hairy beneath, the outer not radiating; styles with enlarged base, forming the stylopodium. *Fruit* 2–2.5 mm, ovoid, laterally compressed; commissure broad; mericarps with slender ridges paler than the reddish-brown vittae; carpophore present; pedicels c. 5 mm, glabrous; styles shorter than the petals in flower and usually diverging, elongating later but often breaking off as the fruit ripens; stigma capitate. *Cotyledons* abruptly contracted into a petiole. $2n = 36, 40$. Flowering in July and August.

In grassy and rocky places throughout most of the British Isles, except for parts of the north and west; commonest on chalk and limestone. Most of Europe.

Fruit × 12; fruit t.s. × 12.

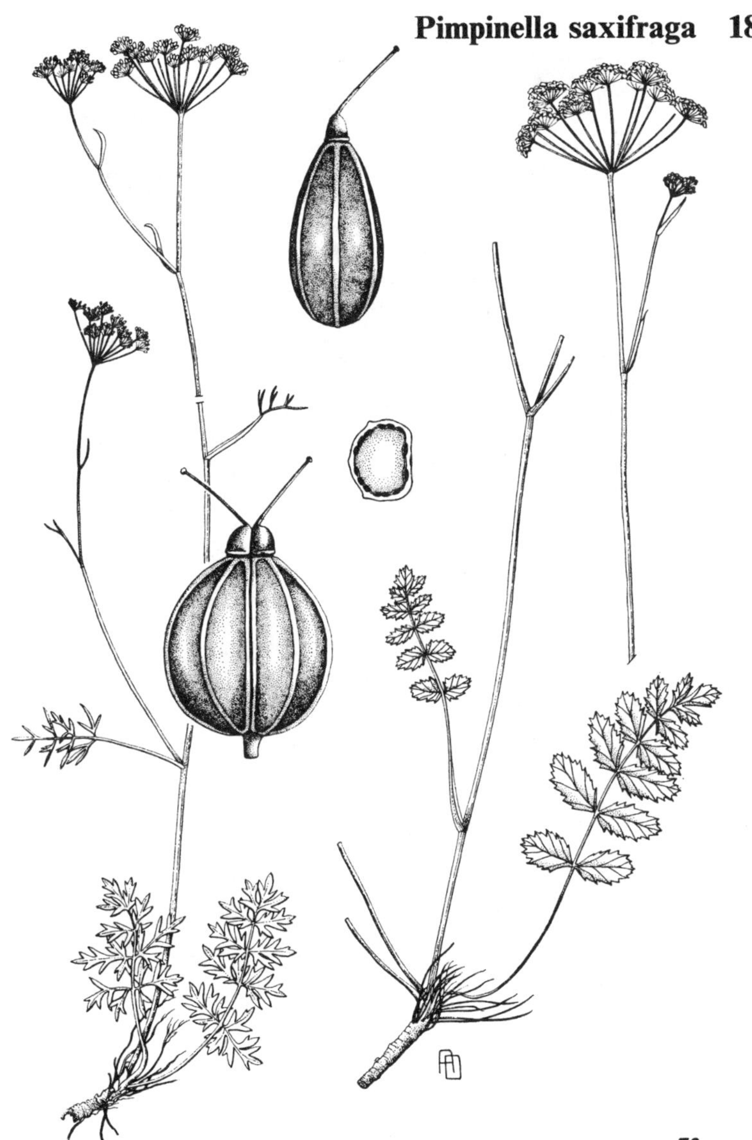

19. **Aegopodium podagraria** L.

Ground-Elder, Goutweed

A glabrous perennial with long slender rhizomes. *Stems* up to 100 cm, hollow, grooved. *Leaves* 1- to 2-ternate, the lobes up to c. 80 × 40 mm, ovate, often scabrid on the veins, often unequal at the base, serrate, the teeth cartilaginous; lower leaves long-petiolate, the upper much smaller and with short, sheathing petioles. *Umbels* compound, hemispherical, with 10–21 smooth rays 1–4 cm long; peduncle longer than the rays; umbels with most flowers hermaphrodite. *Bracts* and bracteoles usually absent. *Flowers* white; sepals absent; outer petals somewhat larger than the inner; styles with enlarged base, forming the stylopodium. *Fruit* 3–4 mm, ovoid, laterally compressed; commissure broad; mericarps with slender ridges; vittae absent when the fruit is mature; carpophore present; pedicels 3–8 mm, glabrous; styles short and erect in flower, elongating and becoming appressed to the mature fruit; stigma capitate. *Cotyledons* tapering gradually at the base, without a distinct petiole. $2n = $?22, 42. Flowering in May and June.

Perhaps native in some localities but usually naturalized from gardens, where it was formerly grown as a pot-herb. Grassy places near buildings, on roadsides and an extremely persistent garden weed throughout most of the British Isles. A native woodland plant in most of Europe and temperate Asia.

Easily recognized by the long slender rhizomes and ternate leaves with large lobes.

The genus contains 5–7 species.

Fruit × 10; fruit t.s. × 10.

20. **Sium latifolium** L.

Greater Water-Parsnip

A glabrous perennial. *Stems* up to 200 × 2 cm, hollow, grooved. *Submerged leaves* 2- to 3-pinnate, with linear lobes, present only in spring. *Aerial leaves* simply pinnate, with 3–6(–16) pairs of lobes; lobes up to 12 × 4 cm, lanceolate to ovate, unequal at the base, serrate, with cartilaginous margin and teeth; lower leaves long-petiolate, the upper with short sheathing petioles. *Umbels* compound, with usually 20–30 smooth rays 1.5–4 cm long; peduncle shorter to longer than the rays, often leaf-opposed, smooth; terminal umbel with hermaphrodite flowers, the lateral umbels with almost entirely male flowers. *Bracts* usually 2–6, sometimes large and leaf-like; bracteoles usually 3–6, lanceolate. *Flowers* white; sepals linear-lanceolate; petals papillose beneath, the outer not radiating; styles with enlarged base, forming the stylopodium. *Fruit* 3–4 mm, ovoid, laterally compressed, constricted at the commissure; mericarps with thick prominent ridges and 3 superficial vittae; carpophore present; pedicels 5–10 mm; styles very short and stout, recurved in fruit; stigma spathulate. *Cotyledons* abruptly contracted into a petiole. $2n = 20$. Flowering in July and August.

In shallow water, especially in fen-ditches, very local and mainly to the S.E. of a line from the Humber to the Bristol Channel; apparently decreasing. Most of Europe; Siberia.

There are 10–15 species of *Sium* distributed throughout most of the world except for South America and Australia. One, *S. sisarum* L., is sometimes cultivated for its tuberous roots, which are edible.

Fruit × 8; fruit t.s. × 5.

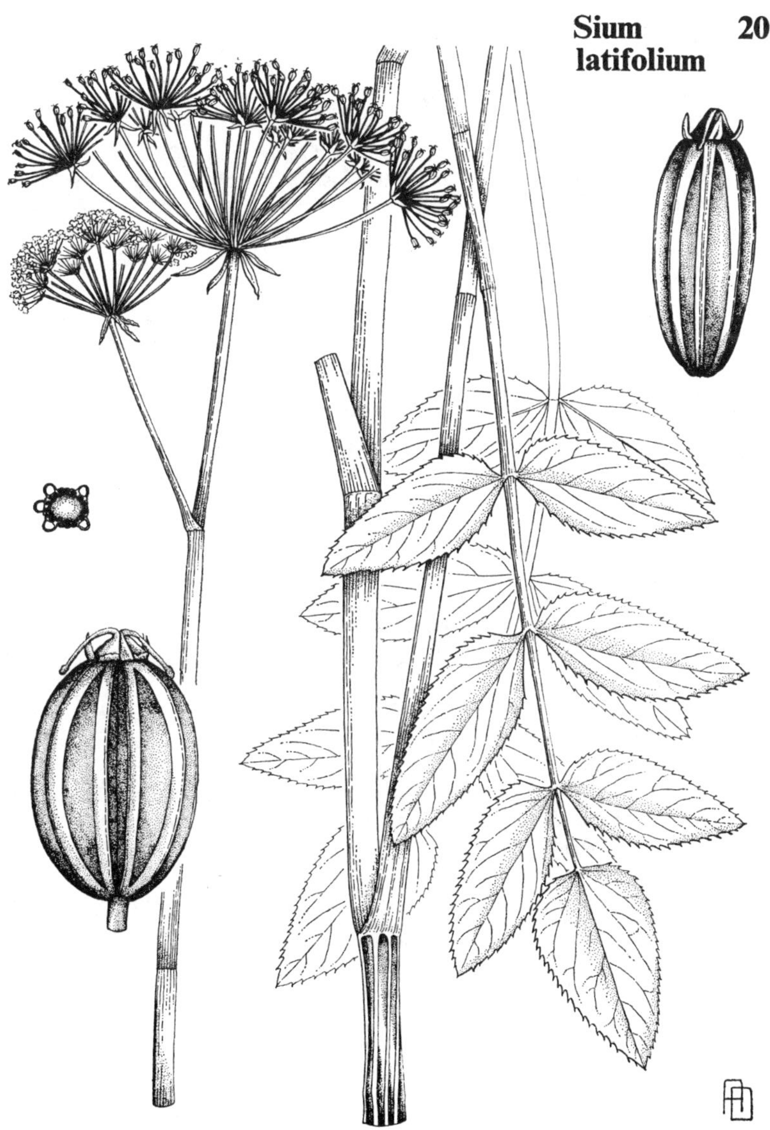

Sium
latifolium
20

21. **Berula erecta**   (Hudson) Coville

Lesser Water-Parsnip

A glabrous, stoloniferous perennial. *Stems* up to 100 × 1 cm, erect or decumbent, hollow, grooved. *Leaves* simply pinnate, the lower with 5–9(–14) pairs of lobes; lobes up to 5 × 2 cm, oblong-lanceolate to ovate, cuneate to rounded and somewhat unequal at the base, 1- to 2-serrate, sometimes lobed, the teeth acute, cartilaginous or sometimes rounded with a cartilaginous mucro; lower leaves long-petiolate, the upper with short sheathing petioles. *Umbels* compound, with usually 7–18 smooth rays 0.5–2.5 cm long; peduncle as long as or somewhat longer than the rays, usually leaf-opposed; umbels all with hermaphrodite flowers. *Bracts* usually 4–7, lanceolate, 3-fid or pinnatifid; bracteoles usually 4–7, lanceolate or 3-fid. *Flowers* white; sepals triangular, sometimes unequal; petals smooth beneath, the outer not radiating; styles with enlarged base, forming the stylopodium. *Fruit* 1.5–2 mm, globose, somewhat compressed laterally, constricted at the commissure; mericarps with slender ridges and vittae sunk in the pericarp; carpophore present; pedicels 3–5 mm, smooth; styles recurved, little longer than the stylopodium; stigma capitate. *Cotyledons* abruptly contracted into a petiole. $2n = 20$. Flowering from July to September.

In damp places and shallow water. Fairly common in lowland districts of England, except the south-west; rare in most of Wales and Scotland; frequent in C. Ireland. Most of Europe and W. Asia.

B. erecta is often confused with *Apium nodiflorum* but is readily distinguished from it by the presence of 4–7 bracts and by having a distinct 'node' in the lower part of the petiole of the basal leaves.

The genus contains about 3 species in Europe, temperate Asia and North America.

Fruit × 15; fruit t.s. × 10.

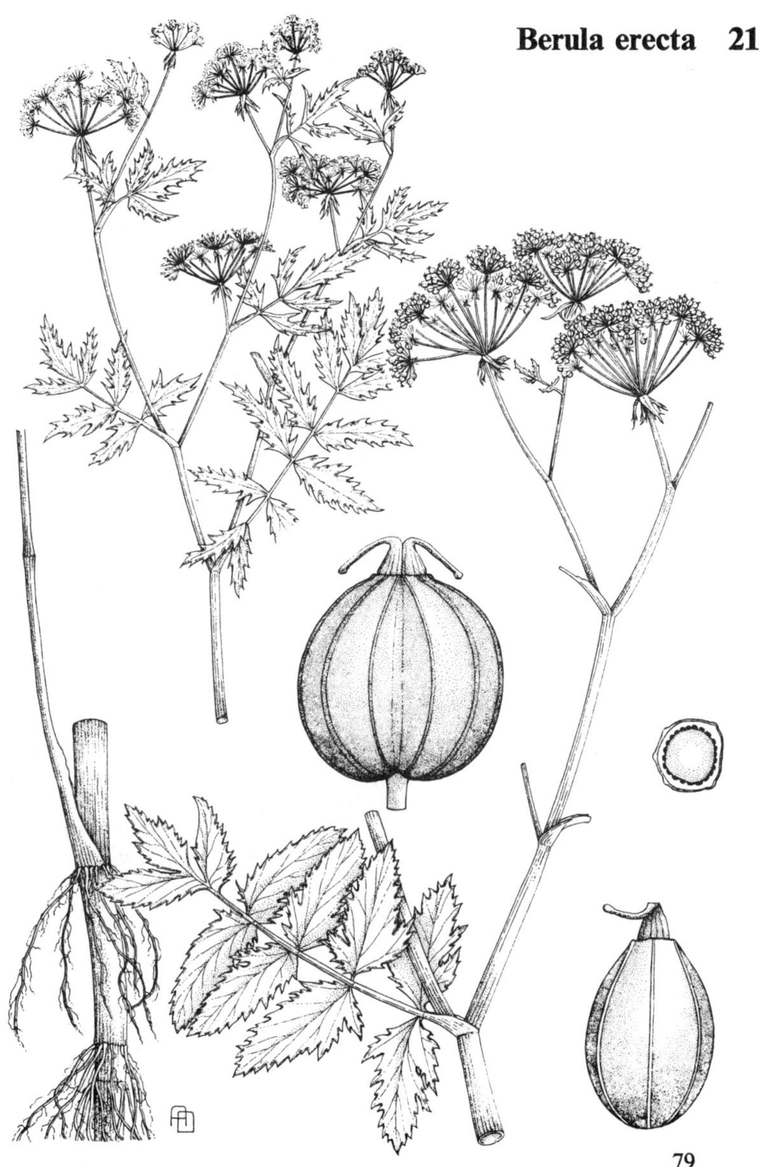

22. **Crithmum maritimum** L.

Rock Samphire

A glabrous perennial. *Stems* up to 45 cm, woody below, solid, striate. *Leaves* 2- to 3-pinnate, fleshy, deltate in outline; lobes (10–) 20–50 mm, subulate to linear-lanceolate, acute, entire; lower leaves with long petioles, shortly sheathing at the base, the upper with short, entirely sheathing petioles; upper leaves not much smaller than the lower. *Umbels* compound, with 8–36 smooth rays 1–4 cm long, thickening in fruit; peduncles usually longer than the rays; all umbels with hermaphrodite flowers. *Bracts* 5–10, lanceolate, broadly membranous, ultimately deflexed; bracteoles 6–8, lanceolate, deflexed in fruit. *Flowers* yellowish-green; sepals very small; outer petals not radiating; styles with enlarged base, forming the stylopodium. *Fruit* 5–6 mm, ovoid-oblong, not compressed; commissure broad; mericarps with thick prominent ridges; vittae several in each groove of the spongy mesocarp; carpophore present; pedicels 5–10 mm; styles shorter than the stylopodium, divergent; stigma truncate. *Cotyledons* tapered gradually at the base, without a distinct petiole. $2n = 20$. Flowering from June to August.

On sea-cliffs and rocks, more rarely on sand or shingle near the sea. Common on the south and west coasts of England and Wales and in Ireland; rare in Scotland and only in the west. Coasts of the N. Atlantic (excluding North America), the Mediterranean and the Black Sea.

The plant when crushed has a strong characteristic smell resembling that of furniture polish. It was formerly used as a pickle. Samuel Pepys records being given a barrel of samphire on 21 September 1660 and J. D. Hooker refers to 'the well-known pickled condiment'. It was also well-known to Shakespeare, who in *King Lear* has the lines:

> . . . ; half-way down
> Hangs one that gathers samphire, dreadful trade!

There is only one species in the genus.

Fruit × 5; fruit t.s. × 5.

Crithmum 22
maritimum
81

23. **Seseli libanotis** (L.) Koch

Moon Carrot

A more or less puberulent biennial or monocarpic perennial. *Stems* up to c. 100 cm, solid, strongly ridged, surrounded at the base by numerous fibrous remains of petioles. *Leaves* usually 2-pinnate, oblong to lanceolate in outline, the pinnae sessile, so that the first pair of lobes is at the base of each pinna; lobes 5–15 mm, lanceolate to ovate in outline, deeply serrate to pinnatifid; basal leaves long-petiolate, the cauline with short, inflated sheathing petioles. *Umbels* compound, with (10–)20–60 more or less puberulent rays, usually 2–3 cm long; peduncle longer than the rays, more or less hairy, at least at the top; all umbels with hermaphrodite flowers. *Bracts* usually 8 or more, linear, erect to more or less deflexed; bracteoles 10–15. *Flowers* white; sepals conspicuous, deciduous; outer petals not radiating; styles with enlarged base, forming the stylopodium. *Fruit* 2.5–3 mm, ovoid, scarcely compressed, densely to rather sparsely puberulent; commissure broad; mericarps with broad ridges; carpophore present; vittae solitary; pedicels 2–7 mm, hairy; styles rather longer than the stylopodium, divergent to recurved; stigma capitate. *Cotyledons* abruptly contracted into a petiole. $2n = 22$. Flowering in July and August.

Rough grassy and bushy places on chalk, very local. E. Sussex, Hertfordshire, Bedfordshire and Cambridgeshire. Most of Europe, W. Asia and North Africa.

Represented in Britain by subsp. **libanotis** which occurs mainly in W. and C. Europe.

Very variable in leaf-dissection and hairiness, but always easily recognized by the dense fibres at the stem-base and the hairy ovary and fruit. The petals are said to be sometimes hairy beneath, but in all the British material examined they are glabrous.

This is our only representative of a genus of c. 80 species occurring mainly in Europe and temperate Asia.

Fruit $\times$ 10; fruit t.s. $\times$ 10.

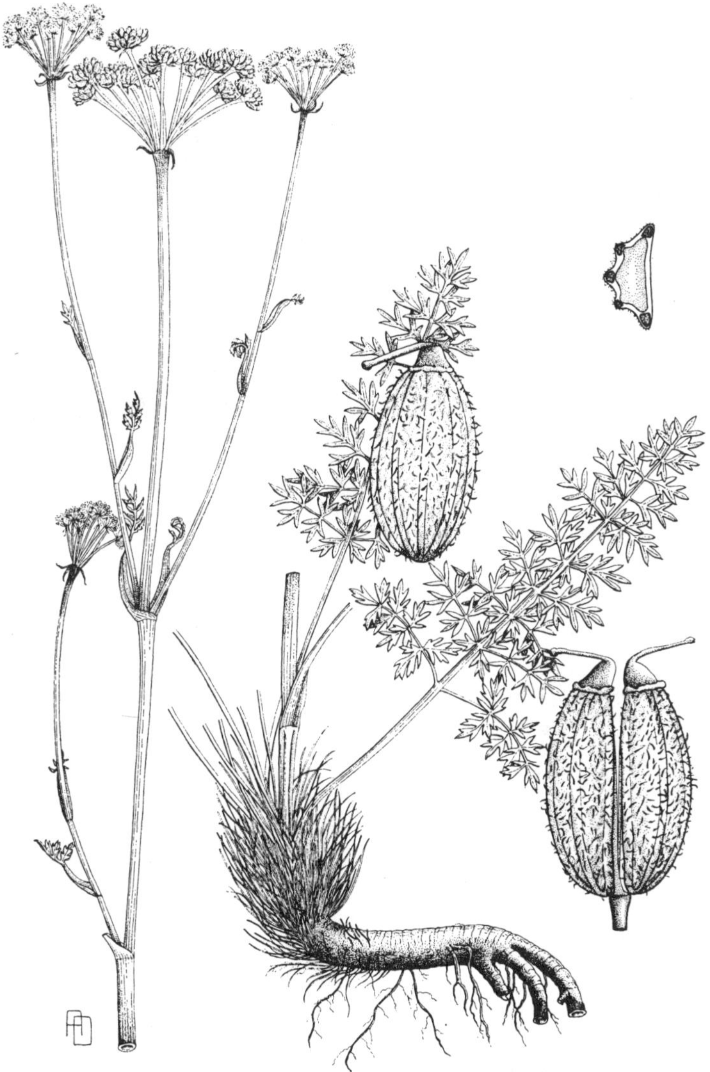

24. **Oenanthe fistulosa** L.

Tubular Water-Dropwort

A glabrous, stoloniferous perennial. *Roots* tuberous, fusiform. *Stems* up to 80 × c. 0.5 cm, fistular, often constricted at the nodes, striate. *Leaves* 1- to 2(–3)-pinnate, oblong to lanceolate in outline, the lobes linear to lanceolate; basal leaves 2-pinnate, sometimes submerged, soon withering, the cauline mostly 1-pinnate with entire lobes 0.5–2 cm; petioles of cauline leaves fistular, longer than the blade. *Umbels* compound, with 2–4 rays usually 1–3 cm long, thickening after flowering; peduncle longer than the rays; terminal umbels with hermaphrodite and some male flowers, the lateral umbels with male flowers. *Bracts* absent; bracteoles 7–16, linear. *Partial umbels* globose in fruit, the pedicels thickening after flowering. *Flowers* white or pinkish; sepals conspicuous, acute, persistent; outer petals somewhat radiating; styles with enlarged base, forming the stylopodium. *Fruit* 3–4 mm, obconical to cylindrical; commissure broad; mericarps with inflated corky ridges; carpophore present; vittae solitary; styles at least as long as the fruit, erect; stigma a small knob. *Cotyledons* abruptly contracted into a petiole. $2n = 22^*$. Flowering from July to September.

Marshy places and shallow water, mainly in the eastern half of England, very local in Scotland and Wales and mainly in the eastern half of Ireland. Most of Europe, W. Asia and N.W. Africa.

The genus contains about 35 species in Europe, temperate Asia and North Africa.

Fruit × 5; fruit t.s. × 4.

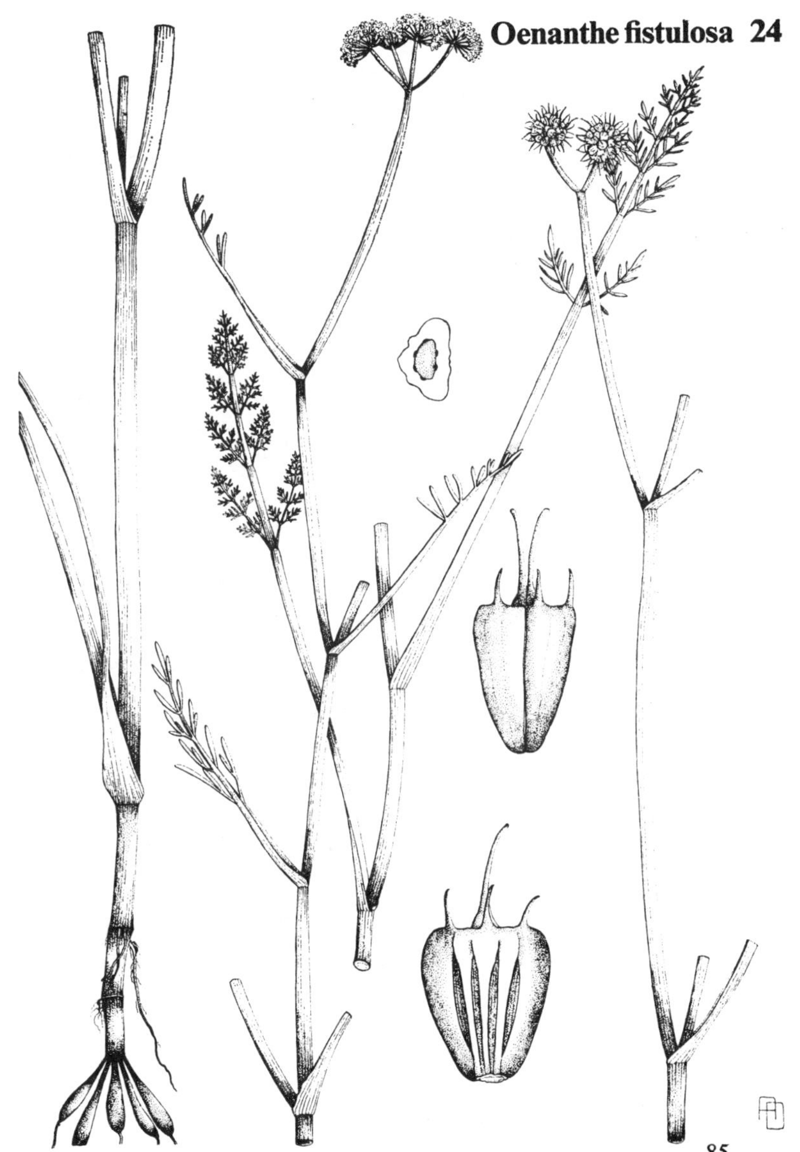

Oenanthe fistulosa 24
85

25. **Oenanthe pimpinelloides** L.

Corky-fruited Water-Dropwort

A glabrous perennial. *Roots* with ovoid tubers distant from the base of the stem. *Stems* up to 100 × 0.5 cm, solid, strongly grooved. *Lower leaves* 2-pinnate, long-petiolate, the lobes c. 5 mm, lanceolate to ovate, cuneate at the base, deeply toothed or pinnatifid; *upper leaves* 1- to 2-pinnate, the blade at least as long as the petiole, and the lobes 10–30 mm, linear, entire. *Umbels* compound, with 6–15 smooth rays 1–2 cm long, thickening after flowering; peduncle longer than the rays; terminal umbels with long-pedicellate male flowers and shortly pedicellate hermaphrodite flowers, the lateral umbels with male flowers. *Bracts* 1–5, linear to linear-lanceolate; bracteoles 12–20, linear to linear-lanceolate. *Partial–umbels* flat-topped in fruit, the pedicels thickening after flowering, especially near their glabrous apex. *Flowers* white; sepals conspicuous, acute, persistent; outer petals somewhat radiating; styles with enlarged base, forming the stylopodium. *Fruit* c. 3.5 mm, cylindrical; commissure broad; mericarps with prominent ridges; carpophore present; vittae solitary; styles about as long as the fruit, erect; stigma a small knob. *Cotyledons* abruptly contracted into a petiole. $2n = 22$. Flowering in June and July.

In damp meadows and other moist grassy places. Locally common from E. Devon and N. Somerset to Hampshire, very local elsewhere south of a line from Worcestershire to Essex; formerly in one locality in Co. Cork. W. and S. Europe, S.W. Asia.

Fruit × 7; fruit t.s. × 7.

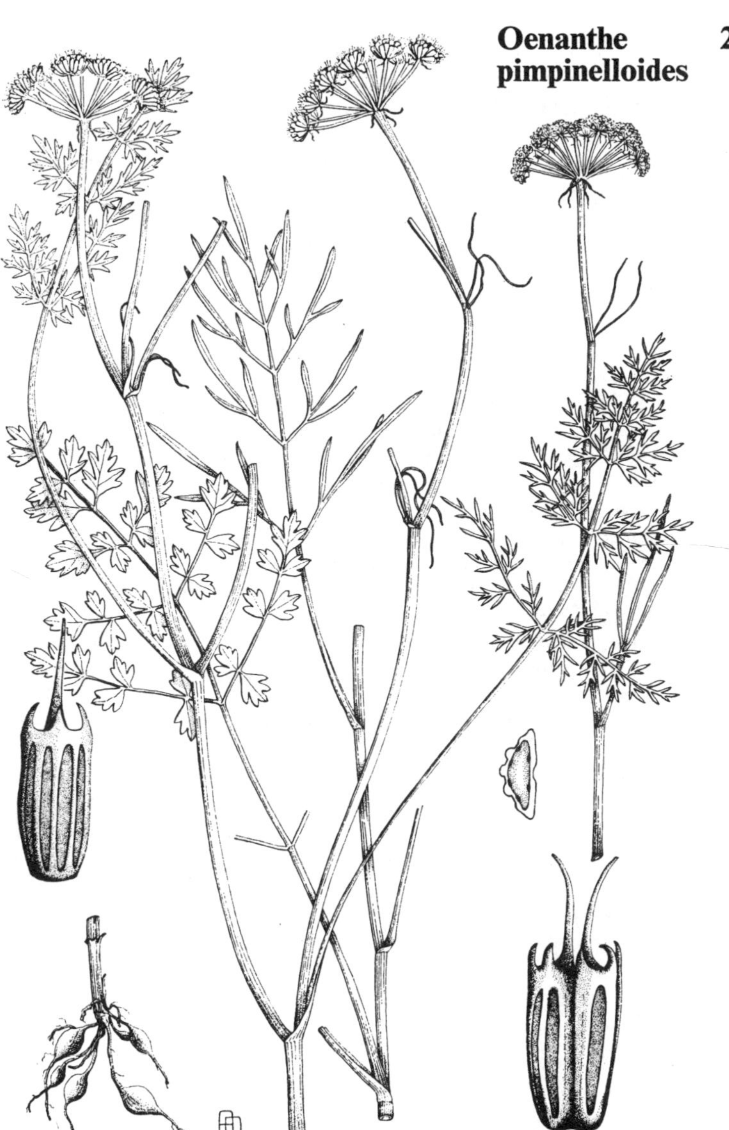

26. **Oenanthe silaifolia** Bieb.

Narrow-leaved Water-Dropwort

A glabrous perennial. *Roots* with obovoid or fusiform tubers which taper towards their junction with the stem. *Stems* up to 100 × 0.7 cm, solid at the base, hollow above, grooved and striate. *Lower leaves* 2- to 4-pinnate, long-petiolate, soon withering, the lobes usually 10–30 mm, linear to linear-lanceolate; *upper leaves* 1- to 2-pinnate, with linear-lanceolate lobes, the petiole shorter than the blade. *Umbels* compound, with 4–8(–10) smooth rays 1.5–3 cm long, thickening after flowering; peduncle longer than the rays; terminal umbels with long-pedicellate male flowers and shortly pedicellate hermaphrodite flowers, the lateral umbels with male flowers. *Bracts* usually absent; bracteoles 10–17, lanceolate, acute. *Partial umbels* not flat-topped in fruit, the pedicels thickening after flowering. *Flowers* white; sepals conspicuous, acute, persistent; outer petals somewhat radiating; styles with enlarged base, forming the stylopodium. *Fruit* 3–3.5 mm, cylindrical; commissure broad; mericarps with prominent ridges; carpophore present; vittae solitary; styles shorter than the fruit, erect to somewhat divergent; stigma tapering. *Cotyledons* abruptly contracted into a petiole. $2n = 22^*$. Flowering in June.

Wet meadows, usually near rivers, very local and apparently decreasing. South and east of a line from the Severn to the Wash with outlying stations in Worcestershire and Nottinghamshire. W., C. and S. Europe, S.W. Asia, N.W. Africa.

Similar to *O. pimpinelloides* when in flower, but easily distinguished by all the leaves having linear to linear-lanceolate, entire lobes, while the lower leaves of *O. pimpinelloides* have lanceolate to ovate, toothed or pinnatifid lobes.

Fruit × 7: fruit t.s. × 7.

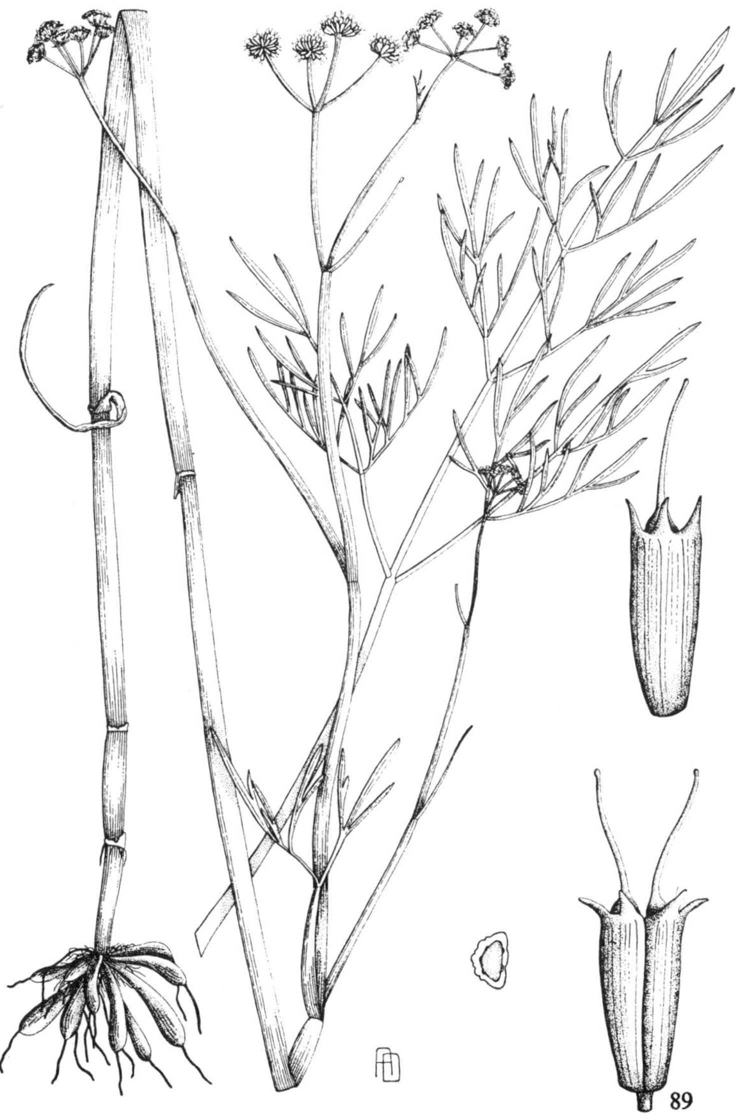

89

27. **Oenanthe lachenalii** C. C. Gmelin

Parsley Water-Dropwort

A glabrous perennial. *Roots* tuberous, cylindrical or fusiform. *Stems* up to 100 × c. 0.4 cm, solid, sometimes developing a small cavity when old, striate. *Lower leaves* (1–) 2(–3)-pinnate, long-petiolate, soon withering, the lobes usually 10–20 mm, linear to spathulate or rarely narrowly obovate, entire or sometimes pinnatifid; *upper leaves* 1- to 2-pinnate, with linear to linear-lanceolate lobes usually 15–50 mm, the petiole shorter than the blade. *Umbels* compound, with 5–9(–20) smooth rays usually 1–3 cm long, not thickening after flowering; peduncle longer than the rays; umbels all with male and hermaphrodite flowers, the proportion of the latter decreasing in the later lateral umbels. *Bracts* up to c. 5, subulate or linear-lanceolate; bracteoles usually 5–7, oblong-lanceolate, acute. *Partial umbels* not flat-topped in fruit; pedicels not thickening after flowering. *Flowers* white; sepals conspicuous, acute, persistent; outer petals somewhat radiating; styles with enlarged base, forming the stylopodium. *Fruit* c. 2.5 mm, ovoid; commissure broad; mericarps with prominent slender ridges; carpophore present; vittae solitary; styles shorter than the fruit, divergent or recurved; stigma tapering. *Cotyledons* abruptly contracted into a petiole. $2n = 22^*$. Flowering from June to September.

In marshes and fens, often near the coast and in somewhat brackish places. Much of the British Isles, but absent from N. Scotland and many inland areas. W. Europe, extending eastwards to Poland and Jugoslavia; Algeria (very rare).

Fruit × 10.

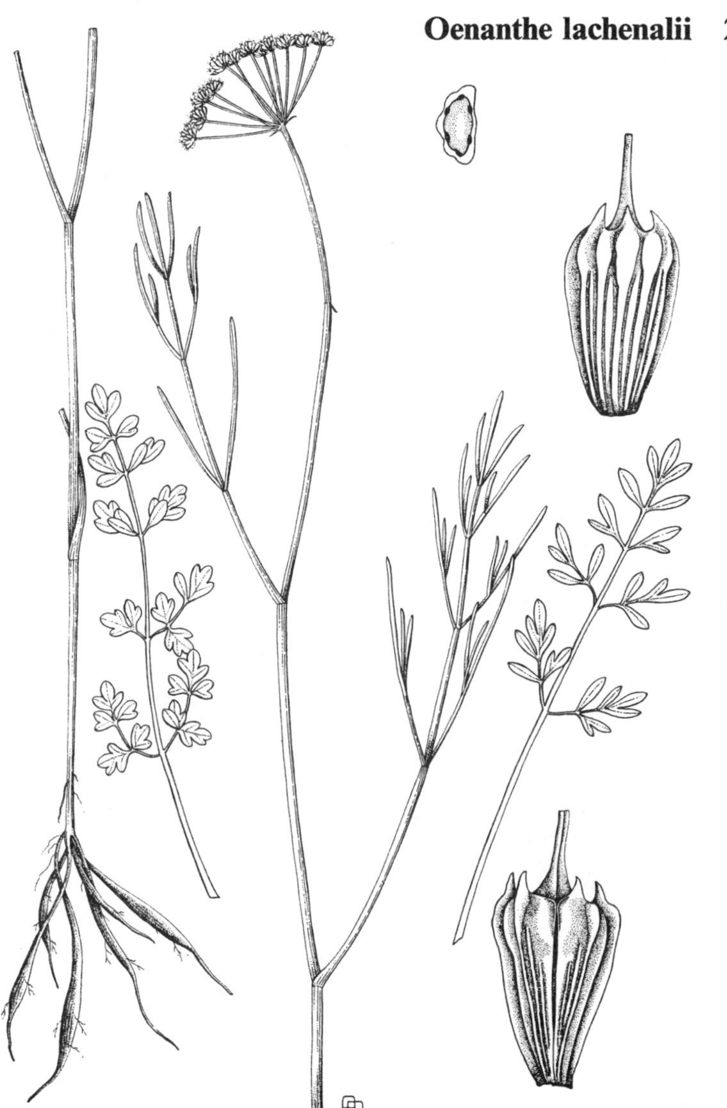

28. **Oenanthe crocata** L.

Hemlock Water-Dropwort

A glabrous perennial. *Roots* with cylindrical-obovoid tubers up to c. 6 × 1 cm. *Stems* up to 150 × 1 cm, hollow, grooved. *Lower leaves* 3(–4)-pinnate, with mostly sheathing petioles, the lobes usually 10–20 mm, ovate to suborbicular in outline, crenate to pinnatifid, usually cuneate at the base; *upper leaves* 1- to 2-pinnate, usually with narrower lobes and a short, sheathing petiole. *Umbels* compound, with (7–)12–40 smooth rays (1.5–)3–8 cm long, not thickening after flowering; peduncle longer than the rays; terminal umbels with mostly hermaphrodite flowers, the lateral with largely or wholly male flowers. *Bracts* c. 5, linear to 3-fid; bracteoles 6 or more. *Partial umbels* not flat-topped in fruit, the pedicels not thickening after flowering. *Flowers* white; sepals conspicuous, ovate to triangular, acute, persistent; outer petals scarcely radiating; styles with enlarged base, forming the stylopodium. *Fruit* 4–5.5 mm, cylindrical, rarely subovoid; commissure broad; mericarps with slender ridges; carpophore present; vittae solitary; styles about half as long as the fruit; stigma a small knob. *Cotyledons* abruptly contracted into a petiole. $2n = 22^*$. Flowering in June and July.

Wet places; usually calcifuge. Mainly in the south and west of Great Britain; in Ireland absent from much of the centre and west. W. Europe, N.W. Africa.

The tubers are sweetish-tasting, but very poisonous, due to a series of polyacetylenes. The active principle is oenanthetoxin, a convulsant poison, which can cause rapid death with few symptoms. Fatal cases of human poisoning have occurred when the leaves were mistaken for those of celery, or the tubers for parsnips; cattle poisoning is not uncommon.

Fruit × 7; fruit t.s. × 7.

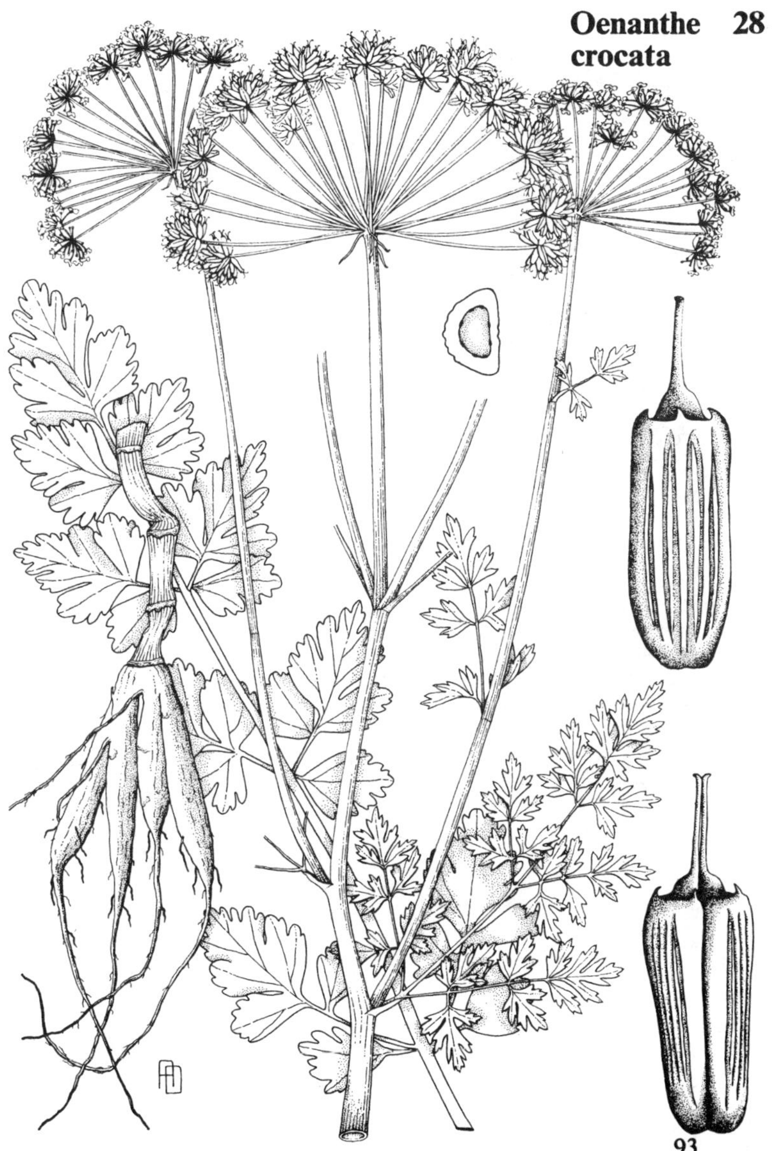

93

29. **Oenanthe fluviatilis** (Bab.) Coleman

River Water-Dropwort

A glabrous perennial. *Roots* of mature plants not tuberous. *Stems* up to 100 × 1 cm, hollow, striate, mostly submerged and ascending or erect. *Lower submerged leaves* 2-pinnate, with cuneate lobes deeply cut into linear to filiform segments; *aerial leaves* 1- to 3-pinnate, the lobes c. 10 mm, ovate to suborbicular, shallowly lobed, cuneate at the base; petiole with a sheathing base. *Umbels* compound, leaf-opposed and terminal, with 5–10(–15) smooth or slightly scabrid rays 1–3 cm long, not thickening after flowering; peduncle usually shorter than the rays; most umbels with male and hermaphrodite flowers. *Bracts* usually absent; bracteoles usually 5–8, linear-lanceolate. *Partial umbels* not flat-topped in fruit; pedicels not thickening in fruit. *Flowers* white; sepals conspicuous, lanceolate, acute, persistent; outer petals scarcely radiating; styles with enlarged base, forming the stylopodium. *Fruit* 5–6.5 mm, cylindrical; commissure broad; mericarps with prominent slender ridges; carpophore present; vittae solitary; styles less than $\frac{1}{4}$ as long as the fruit, divergent; stigma a small knob. $2n = 22^*$. Flowering from July to September.

In slowly flowing water. S.E. England, E. and C. Ireland; one locality in Wales; local. Confined to W. Europe.

Young plants may well have tuberous roots which disappear as the stem elongates and fibrous roots develop at the nodes, as appears to happen in *O. aquatica*. Investigation in the field is needed. Vegetative spread by detached fragments, which root readily, is frequent.

Fruit × 7; fruit t.s. × 7.

30. **Oenanthe aquatica** (L.) Poiret

Fine-leaved Water-Dropwort

A robust glabrous annual or biennial. *Roots* of young plants tuberous liké those of *O. crocata*, tubers disappearing as the plant flowers. *Stems* up to 150 × 1 cm, hollow, striate. *Lower leaves* 3- to 4-pinnate, finely divided, with filiform lobes when submerged; *aerial leaves* 2- to 3-pinnate, the lobes c. 5 mm, lanceolate to ovate in outline, pinnatifid, acute; petiole with a sheathing base. *Umbels* compound, leaf-opposed and terminal, with (4–)6–16 usually scabrid rays 1–4 cm long, not thickening after flowering; peduncle usually shorter than the rays; most umbels with male and herma-phrodite flowers. *Bracts* usually absent; bracteoles usually 4–8, linear-lanceolate. *Partial umbels* not flat-topped in fruit, the pedicels not thickening in fruit. *Flowers* white; sepals conspicuous, narrowly triangular, acute, persistent; outer petals scarcely radiating; styles with enlarged base, forming the stylopodium. *Fruit* 3.5–4.5 mm, ovoid; commissure broad; mericarps with prominent, rather thick ridges; carpophore present; vittae solitary; styles less than $\frac{1}{4}$ as long as the fruit, divergent; stigma a small knob. $2n = 22^*$. Flowering from June to September.

In still or slowly flowing water, often in places that dry up in summer. England and E. Wales; S.E. Scotland; C. Ireland. Most of Europe extending eastwards to C. Asia.

Fruit × 7; fruit t.s. × 7.

97

31. **Aethusa cynapium** L.

Fool's Parsley

A glabrous annual. *Stems* 5–120 cm, hollow, finely striate. *Leaves* 2- to 3-pinnate, triangular to lanceolate in outline, the lower withered at flowering; lobes usually 5–15 mm, pinnatifid, lanceolate to ovate in outline, acute, with antrorsely scabrid margins; petiole short, mostly sheathing. *Umbels* compound, leaf-opposed and terminal with 4–20 rays 0.5–3 cm long, which are antrorsely serrulate on the angles; peduncles longer than the rays; umbels with almost entirely hermaphrodite flowers. *Bracts* usually absent; bracteoles usually 3-4, on the outer side of the partial umbels, deflexed, linear, long-acute. *Flowers* white; sepals absent; outer petals scarcely radiating; styles with enlarged base, forming the stylopodium. *Fruit* 2.5–3.5 mm, broadly ovoid, somewhat compressed dorsally; commissure broad; mericarps with prominent, keeled ridges; carpophore present; vittae solitary; pedicels slender; styles as long as and closely appressed to the stylopodium; stigma capitate. *Cotyledons* of the seedling abruptly contracted into a petiole. $2n =$ 20. Flowering from June to September.

Cultivated ground and waste places throughout much of the British Isles, but mainly coastal in Scotland. Throughout most of Europe, S.W. Asia and N.W. Africa.

Two subspecies occur in Britain: subsp. **cynapium** with stems usually 30–80 cm, the bracteoles usually several times as long as the longer pedicels and the outer pedicels about twice as long as their fruits, and subsp. **agrestis** (Wallr.) Dostál, with stems usually 5–20 cm, the bracteoles usually about as long as the longer pedicels and the outer pedicels usually shorter than their fruits. The former occurs in gardens, on roadsides etc. and the latter mainly in arable fields. The plant illustrated is subsp. *cynapium.*

A. cynapium is very poisonous, containing polyacetylenes (often erroneously called alkaloids), coniine and cynapine. These do not withstand drying, so that hay containing *Aethusa* is not poisonous to livestock.

The genus has only this one species.

Fruit × 10; fruit t.s. × 7.

32. **Foeniculum vulgare** Miller

Fennel

A glabrous rather glaucous perennial or biennial. *Stems* up to 250 cm, solid, developing a small hollow when old, finely striate. *Leaves* 3- to 4-pinnate, more or less triangular in outline; lobes usually 5–50 mm, filiform, with a cartilaginous apex, usually widely spaced and not all lying in the same plane; petiole with a sheathing base or, in the upper leaves, entirely sheathing. *Umbels* compound, leaf-opposed and terminal, with 10–40 smooth rays 1–6 cm long; peduncles usually longer than the rays; all umbels with hermaphrodite flowers. *Bracts* and bracteoles usually absent. *Flowers* yellow; sepals absent; outer petals not radiating; styles with enlarged base, forming the stylopodium. *Fruit* 4–6 mm, ovoid-oblong, scarcely compressed; commissure broad; mericarps with prominent thick ribs; carpophore present; vittae solitary; pedicels 2–5 mm, slender; styles shorter than the stylopodium, divergent or recurved; stigma capitate. *Cotyledons* tapered gradually at the base, without a distinct petiole. $2n = 22$. Flowering from July to October.

On cliffs, roadsides and waste ground, particularly near the sea. Often cultivated for its leaves, which are used in flavouring, and casual or naturalized from gardens. Doubtfully native in Britain. Most of Europe, but probably native only in the south; Mediterranean region.

Represented in Britain by subsp. **vulgare.**

A compact cultivar with greatly enlarged overlapping leaf-bases, var. *azoricum* (Miller) Thell., is cultivated as a vegetable (finocchio).

The genus contains only one species.

Fruit $\times$ 5; fruit t.s. $\times$ 5.

101

33. **Anethum graveolens** L.

Dill

An erect glabrous, somewhat glaucous annual, smelling strongly when crushed. *Stems* up to c. 60 cm, hollow, striate. *Leaves* 3- to 4-pinnate, triangular in outline, the lobes c. 15 mm, filiform, mucronate; petiole shorter than the lamina, mostly sheathing. *Umbels* compound, with 7–30 subequal smooth rays 2–9 cm long; peduncle longer than the rays; umbels mostly with hermaphrodite flowers. *Bracts* and bracteoles absent. *Flowers* yellow; sepals absent; outer petals not radiating; styles with enlarged base, forming the stylopodium. *Fruit* 4.5–6 mm, elliptical, strongly compressed dorsally, dark brown with a paler wing; commissure broad, mericarps with slender low dorsal ridges and winged marginal ridges; carpophore present; vittae solitary; pedicels 4–10 mm, slender; styles about half as long as the stylopodium, recurved and appressed; stigma capitate. *Cotyledons* tapered gradually at the base, without a distinct petiole. $2n = 22$. Flowering in July and August.

Cultivated for the leaves and fruits, which are used for flavouring pickles etc., sometimes persisting for a few years and a fairly frequent casual on waste ground and rubbish tips. Probably native of S.W. Asia, but widely cultivated and naturalized.

A. graveolens is often confused with *Foeniculum* in the vegetative state, but, being an annual, it is easily uprooted and is usually smaller and more slender. The stem is hollow, whereas in *Foeniculum* it is solid, though developing a small hollow when old.

The genus contains only one or two species.

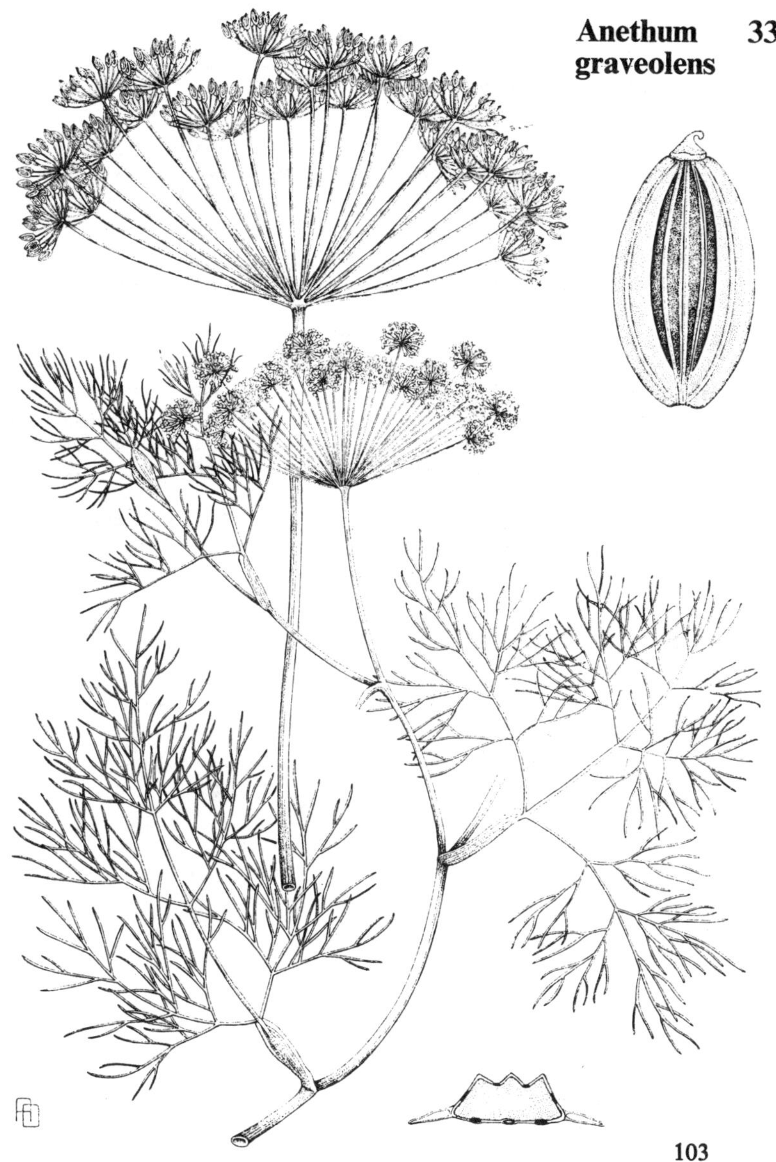

Anethum
graveolens
33

34. **Silaum silaus** (L.) Schinz & Thell.

Pepper-Saxifrage

A glabrous perennial. *Stems* up to c. 100 cm, solid, striate, with a few fibrous remains of petioles at base. *Basal leaves* 2- to 4-pinnate, triangular to lanceolate in outline, long-petiolate, the primary divisions long-stalked; lobes usually 10–15 mm, lanceolate to linear, acuminate or obtuse and mucronate, finely serrulate, with prominent midrib, the apex often reddish; *upper cauline leaves* often 1-pinnate, simple or reduced to sheaths. *Umbels* compound with 4–15 sharply angled rays, the top of the peduncle and the rays papillose; peduncle longer than the rays; umbels mostly with hermaphrodite flowers. *Bracts* 0–3; bracteoles usually 5–11, linear or lanceolate, with scarious margins. *Flowers* yellowish; sepals absent; outer petals not radiating; styles with enlarged base, forming the stylopodium. *Fruit* 4–5 mm, oblong-ovoid, scarcely compressed; commissure broad; mericarps with very prominent, slender ridges, the lateral forming narrow wings; carpophore present; vittae numerous, inconspicuous; pedicels mostly 2–3 mm, rather stout; styles about as long as and appressed to the stylopodium; stigma capitate. *Cotyledons* tapered gradually at the base, without a distinct petiole. $2n = 22^*$. Flowering from June to August.

Meadows and grassy banks. Mainly in S. and E. England, a few localities in Wales and S.E. Scotland; absent from Ireland. W., C. and E. Europe, northwards to the Netherlands and Sweden, but absent from Portugal.

There are about 10 spp. of *Silaum*, mainly in temperate Asia.

Fruit × 6; fruit t.s. × 8.

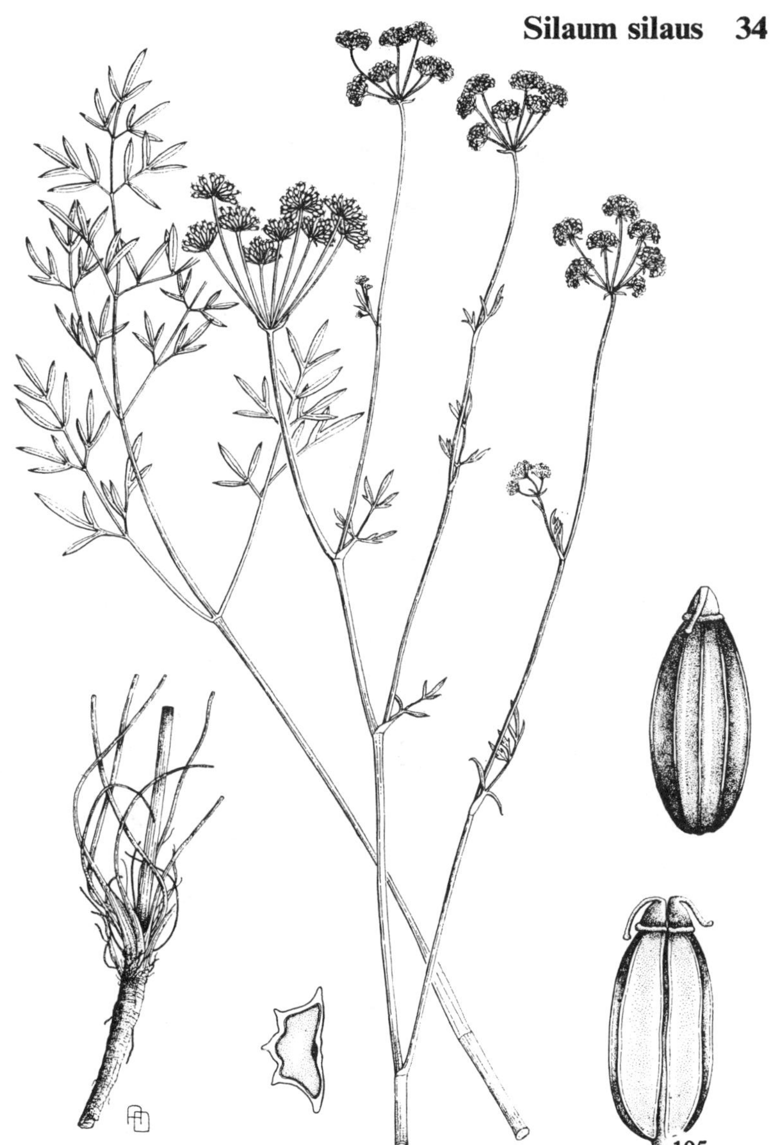

105

35. **Meum athamanticum**   Jacq.

Spignel, Meu, Baldmoney

A glabrous, strongly aromatic perennial. *Stems* up to 60 cm, hollow, striate, surrounded at the base by abundant fibrous remains of petioles. *Basal leaves* 3- to 4-pinnate, lanceolate to ovate in outline, the lobes c. 5 mm, filiform, crowded; petiole slender, abruptly expanded into a largely scarious sheath at base; *cauline leaves* few, small, the petiole usually entirely sheathing. *Umbels* compound, with 6–15 unequal, somewhat scabrid rays 1–5 cm long; peduncle longer than the rays, papillose; terminal umbel with hermaphrodite and a few inner male flowers, the lateral umbels with mostly male flowers. *Bracts* few or sometimes absent, linear to ovate, sometimes lobed, very variable in length; bracteoles usually 3–8, linear, sometimes lobed. *Flowers* white, sometimes pink-tinged; sepals absent; outer petals not radiating; styles with enlarged base, forming the stylopodium. *Fruit* usually c. 7 mm, ovoid, scarcely compressed; commissure broad; mericarps with very prominent thick ridges; carpophore present; vittae 3–5 in each furrow; pedicels 1–7 mm, papillose on the angles; styles little longer than the stylopodium, divergent or somewhat recurved; stigma truncate. $2n = 22$. Flowering in June and July.

Grassland in mountain districts. A few localities in N. England and N. Wales; Scotland northwards to Argyll and Aberdeen, local. W. and C. Europe, extending southwards to the Sierra Nevada and C. Bulgaria.

According to J. D. Hooker the root was formerly eaten in Scotland.

The genus contains only one species.

Fruit × 4; fruit t.s. × 4.

Meum
athamanticum
35
107

36. **Physospermum cornubiense**   (L.) DC.

Bladderseed

An almost glabrous perennial. *Stems* 30–120 cm, solid, striate. *Basal leaves* 2-ternate, broadly rhombic in outline, the primary divisions long-stalked; lobes usually 15–50 mm, usually ovate in outline, with a cuneate base, pinnatifid or incise-serrate, puberulent on margin and larger veins on both surfaces; petiole long, slender; *cauline leaves* few, small, often simple. *Umbels* compound, with 10–14 smooth, slender, usually subequal rays 2–5 cm long; peduncle usually longer than rays; terminal umbel with hermaphrodite flowers, the lateral with male and hermaphrodite flowers. *Bracts* usually 4–7, lanceolate to linear, acute, membranous; bracteoles similar, but usually fewer. *Flowers* white; sepals triangular; outer petals not radiating; styles with enlarged base, forming the stylopodium. *Fruit* 3–4 mm, inflated, broader than long, didymous, dark brown; commissure narrow; mericarps smooth; carpophore present; vittae solitary; pedicels 5–8 mm, smooth; styles about twice as long as the stylopodium, recurved; stigma truncate. Flowering in July and August.

Open woods and scrub, very local. S. Devon and E. Cornwall; one locality in Buckinghamshire, where it may have been introduced. S. Europe and temperate W. Asia.

The genus contains about 4 species.

Fruit × 10; fruit t.s. × 7.

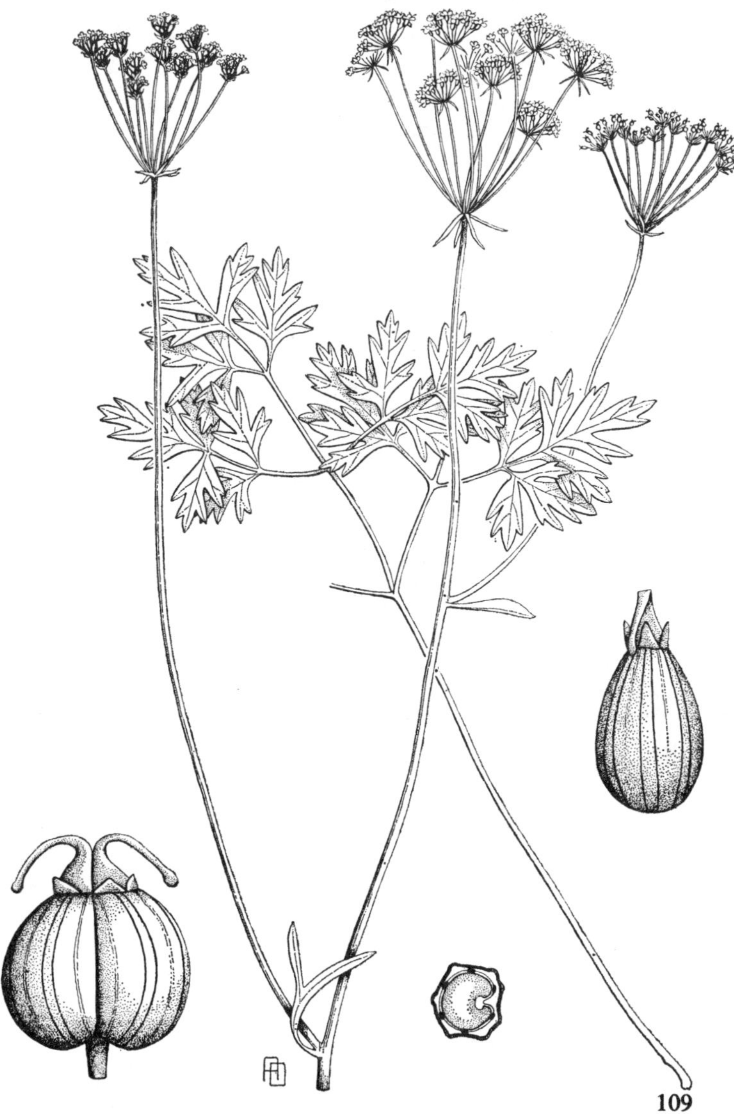

37. **Conium maculatum**   L.

Hemlock

A glabrous biennial. *Stems* up to 200 cm, hollow, striate below, furrowed above, usually pruinose and purple-spotted, at least below. *Basal leaves* 2- to 4-pinnate, triangular in outline, the lobes mostly 10–20 mm, oblong in outline, pinnatifid or incise-serrate, the teeth and lobes with a cartilaginous apex; *upper leaves* similar but smaller; petioles of lower leaves long, with a shortly sheathing base, those of upper leaves short and entirely sheathing. *Umbels* compound, with 10–20 subequal scabrid rays 1–3.5 cm long; peduncle equalling or longer than the rays; terminal umbel with hermaphrodite flowers, the lateral with male and hermaphrodite flowers. *Bracts* 5–6, narrowly triangular to lanceolate, deflexed, with a wide scarious margin; bracteoles 3–6, on the outside of the partial umbel, widened and often connate at the base. *Flowers* white; sepals absent; outer petals not radiating; styles with enlarged base, forming the stylopodium. *Fruit* 2.5–3.5 mm, ovoid, somewhat compressed; commissure narrow; mericarps with prominent, more or less undulate-crenulate ridges; carpophore present; vittae several, slender; pedicels 1–8 mm; styles about as long as the stylopodium, horizontal or recurved; stigma a small knob. *Cotyledons* abruptly contracted into a petiole. $2n = 22$. Flowering in June and July.

In damp places, on roadsides and rubbish tips. Throughout most of the British Isles, but infrequent in the north; Europe and temperate Asia.

Highly poisonous due to the presence of coniine (see *Aethusa*). Readily recognized by the purplish blotching of the stems, which is rarely absent, and by the relatively small umbels and the mousey smell of the plant.

The genus contains this species and one other (but perhaps not congeneric) in S. Africa.

Fruit × 10; fruit t.s. × 10.

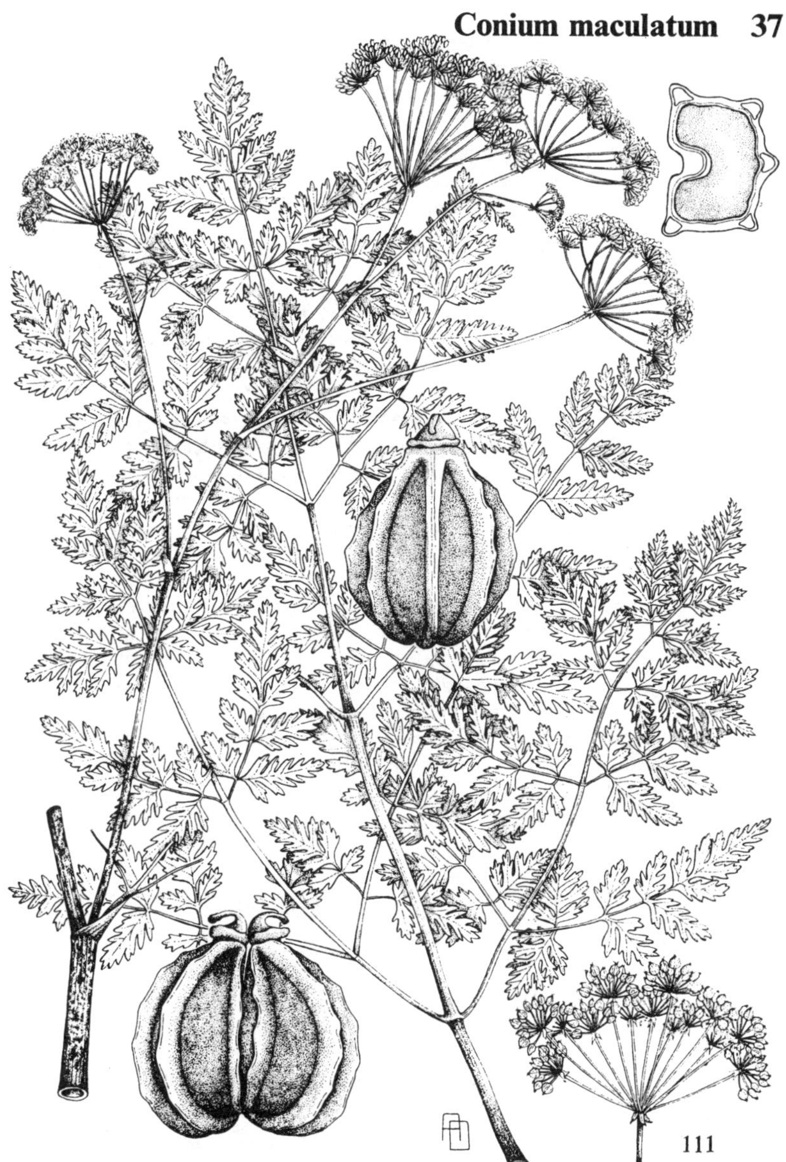

111

38. **Bupleurum rotundifolium** L.

Thorow-Wax

An erect, glabrous, glaucous annual. *Stems* up to c. 30 cm, hollow, smooth, often purple-tinged. *Leaves* 2–5 cm, simple, elliptic-ovate to suborbicular, apiculate, with a cartilaginous margin, the lower attenuate into a petiole, the rest perfoliate; veins numerous, slender, radiating from the base, anastomosing near the margin and connected by slender cross-veins elsewhere. *Umbels* compound, with 4–8 smooth, unequal rays up to 1 cm long; peduncle as long as or a little longer than the rays; all flowers in terminal and lateral umbels hermaphrodite. *Bracts* absent; bracteoles usually 5, 5–12 mm, unequal, oblanceolate to ovate, acuminate, patent in flower, connivent in fruit, shortly connate at base; venation similar to that of the leaf. *Flowers* yellow; sepals absent; outer petals not radiating; styles with an enlarged base, forming the stylopodium. *Fruit* 3–3.5 mm, elliptic-oblong, pruinose, somewhat compressed laterally; commissure broad; mericarps with slender, prominent ridges; carpophore present; vittae inconspicuous; pedicels c. 1 mm; styles 0.2–0.3 mm, much shorter than the stylopodium; stigma truncate. *Cotyledons* tapered gradually at the base, without a distinct petiole. $2n = 16$. Flowering in June and July.

Formerly fairly widespread in the south-eastern half of England, mainly in arable fields and probably dependent on repeated introduction with cereal seed; now very scarce or extinct.

This and the following species are frequently confused.

There are about 150 species of *Bupleurum* in Europe, temperate Asia, North Africa and North America.

Fruit × 7; fruit t.s. × 7.

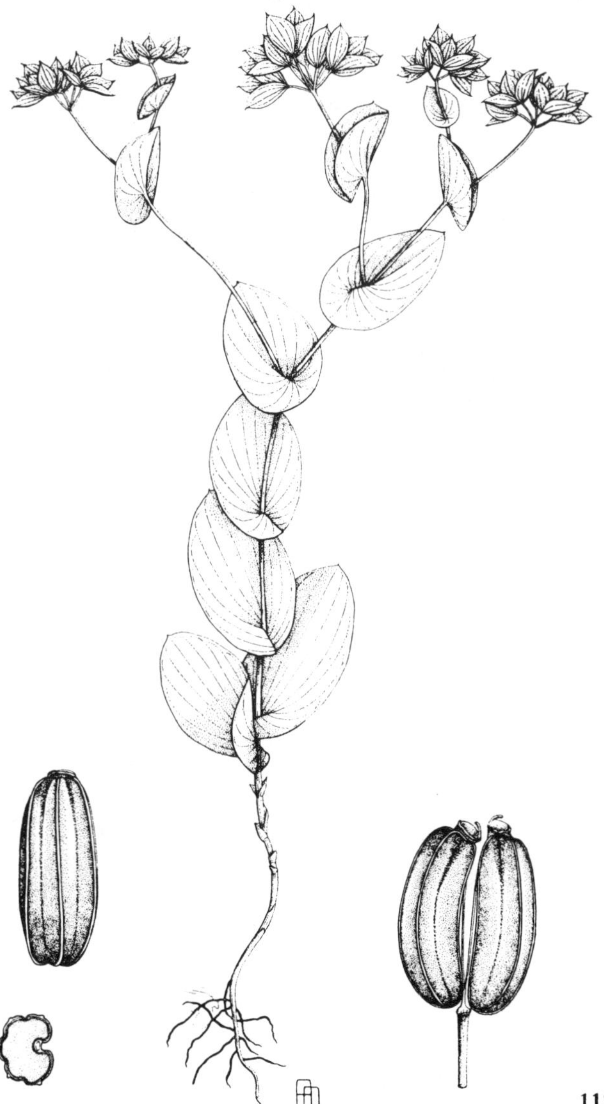

39. **Bupleurum subovatum** Link

False Thorow-Wax

An annual very similar to *B. rotundifolium* from which it differs in having the leaves usually three times as long as wide (instead of about twice as long as wide), the umbels with 2–3 subequal rays and the fruit 4.5–5 mm and strongly tuberculate. The tubercules develop soon after flowering and become more conspicuous as the fruit matures. $2n = 16$. Flowering from June to October.

In Britain a casual in gardens and disturbed ground, where it is introduced by scattering bird seed or with chicken food.

A native of S. Europe, S.W. and C. Asia and N.W. Africa and widely distributed as a casual.

From herbarium specimens it would seem to have been first introduced about 1870, and shows a marked increase in frequency from about 1950 onwards due to the greater availability of seeds for feeding wild birds.

Fruit $\times$ 7; fruit t.s. $\times$ 7.

40. **Bupleurum baldense** Turra

Small Hare's-Ear

A slender, glabrous annual. *Stems* usually less than 10 cm (in Britain), simple or divaricately branched, solid, with raised or very narrowly winged angles. *Leaves* up to 3 cm, simple, linear to narrowly oblanceolate, widest above the middle, acute, sessile, with a somewhat scarious sheathing base; veins 3–5, parallel, without apparent cross veins. *Umbels* compound, with (1–)2–4 unequal rays shorter than the bracts; peduncle longer than the rays; all flowers in terminal and lateral umbels hermaphrodite. *Bracts* c. 4, 5–10 mm, lanceolate, acuminate, glaucous, with 3–5 conspicuous main veins and a narrow scarious margin; bracteoles similar to the bracts but rather smaller and usually 3-veined, concealing the flowers and fruit. *Flowers* yellow; sepals absent; outer petals not radiating; styles with enlarged base, forming the stylopodium. *Fruit* c. 1.5 mm, ellipticoblong, somewhat compressed laterally, often pruinose, commissure broad; mericarps with slender ridges; carpophore present; vittae solitary; pedicels c. 1 mm; styles c. 0.1 mm, horizontal; stigma truncate. *Cotyledons* tapered gradually at the base, without a distinct petiole. $2n = 16$. Flowering in June and July.

In more or less open habitats, such as short turf on rocky ground and grey dunes, very local and always near the sea. Sussex, S. Devon, Channel Islands. W. and S. Europe from England to Romania.

Represented in Britain by subsp. **baldense,** which occurs throughout the range of the species, except for Italy and the Balkan peninsula.

Fruit × 20; fruit t.s. × 12.

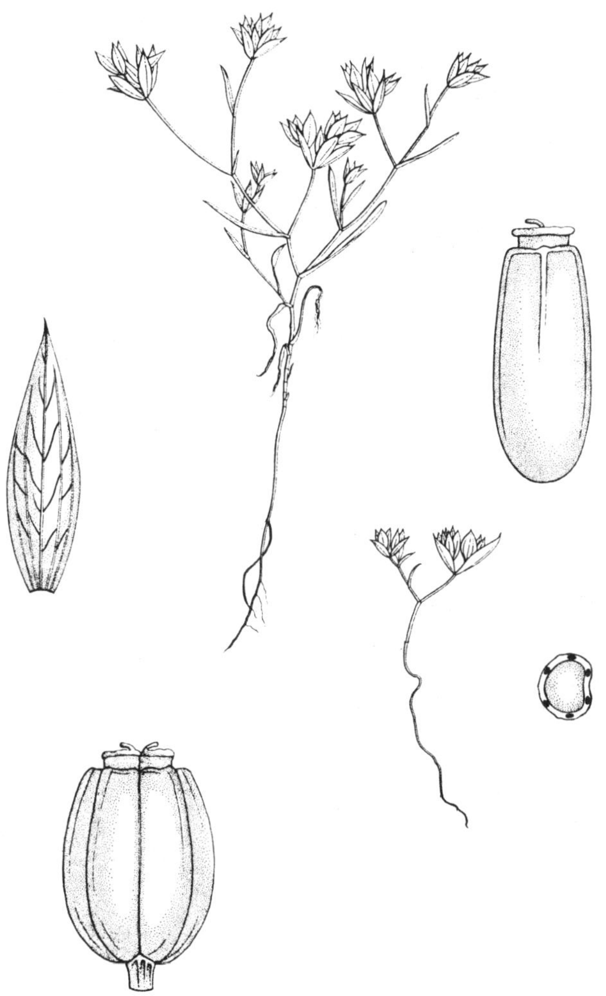

41. **Bupleurum tenuissimum** L.

Slender Hare's-Ear

A glabrous annual. *Stems* up to 50 cm, usually branched and flexuous, solid, striate. *Leaves* up to 7 cm, simple, linear to narrowly oblanceolate, widest above the middle, acute, sessile, scarcely sheathing at the base; veins 3, parallel, with some slender cross-veins. *Umbels* compound, numerous, terminal and axillary, shortly pedunculate or sessile, with (1–)2–3 unequal rays up to 1 cm long; partial umbels with few (often 1–4) flowers; all flowers hermaphrodite. *Bracts* 3–5, up to 6 mm, linear-lanceolate, 3-veined, herbaceous; bracteoles similar to the bracts, not concealing the flowers and fruit. *Flowers* yellow; sepals absent; outer petals not radiating; styles with enlarged base, forming the stylopodium. *Fruit* c. 2 mm, subglobose, somewhat compressed laterally, rugulose-papillose; commissure broad; mericarps with narrowly winged, undulate-crenulate ridges; carpophore present; vittae inconspicuous; pedicels very short; styles c. 0.1 mm, horizontal; stigma truncate. *Cotyledons* tapered gradually at the base, without a distinct petiole. $2n = 16$. Flowering from July to September.

In grassy, somewhat saline habitats. Coast of England from Dorset to the Humber; Bristol Channel; one locality in Co. Durham; sometimes inland, but no recent records. Coasts of most of Europe, extending northwards to Sweden (Gotland); S.W. Asia; N.W. Africa; inland on saline steppe.

Represented in Britain by subsp. **tenuissimum,** which occurs throughout the range of the species except for the south part of the Balkan peninsula and S.E. Russia.

Fruit × 12; fruit t.s. × 8.

42. **Bupleurum falcatum** L.

Sickle-leaved Hare's-Ear

A glabrous perennial with a stout, woody stock. *Stems* up to 100 cm, hollow, striate. *Leaves* simple, the basal c. 10 cm, elliptical, long-petiolate, with 5 parallel veins and a cartilaginous margin; lower cauline linear-lanceolate, often somewhat falcate, narrowed into a petiole, the upper linear, sessile, semiamplexicaul. *Umbels* compound, with 5–11 smooth, usually subequal rays up to c. 3 cm; peduncle longer than the rays; all flowers hermaphrodite. *Bracts* 2–5, very unequal, lanceolate to linear, 3- to 5-veined; bracteoles 4–5, lanceolate, often aristate, usually 3-veined, not concealing the flowers and fruit. *Flowers* yellow; sepals absent; outer petals not radiating; styles with enlarged base, forming the stylopodium. *Fruit* 2.5–3 mm, ellipsoid, somewhat compressed laterally, smooth; commissure broad; mericarps with slender, prominent, narrowly winged ridges; carpophore present; vittae inconspicuous; pedicels 1–3 mm, slender; styles c. 0.5 mm, divergent or appressed to the stylopodium; stigma truncate. *Cotyledons* tapered gradually at the base, without a distinct petiole. $2n = 16$. Flowering from July to October.

First recorded in 1831 and possibly not native; formerly along several miles of road between Ongar and Chelmsford, Essex, now extinct, the last remaining population having been destroyed by hedgerow clearance and ditch cleaning in 1962. S., C. and E. Europe, temperate Asia.

Represented in Britain by subsp. **falcatum,** which occurs throughout the range of the species.

Fruit × 10; fruit t.s. × 10

120

43. **Bupleurum fruticosum** L.

Shrubby Hare's-Ear

A glabrous evergreen shrub. *Stems* up to 250 cm, solid, smooth. *Leaves* up to c. 9 cm, simple, oblanceolate to oblong-obovate, mucronate, pale beneath, sessile, with a cartilaginous margin, a conspicuous, stout midrib, numerous slender, pinnate veins and a close reticulum of fine veins between them. *Umbels* compound, with 3–25 subequal, smooth rays c. 2 cm long; peduncle as long as or longer than the rays; all flowers hermaphrodite. *Bracts* usually 5–6, deciduous, ovate to obovate, with 5–7 longitudinal veins and conspicuous cross-veins; bracteoles 5–6, deciduous, obovate, 4–5 (–7)-veined, not concealing the flowers. *Flowers* yellow; sepals absent; outer petals not radiating; styles with enlarged base, forming the stylopodium. *Fruit* c. 7 mm, oblong-elliptical, somewhat compressed laterally, smooth; commissure broad; mericarps with slender, prominent, narrowly winged ridges; carpophore present; vittae inconspicuous; pedicels c. 5 mm; styles c. 0.3 mm, divergent; stigma truncate. *Cotyledons* gradually narrowed at the base, without a distinct petiole. $2n = 14$. Flowering in July and August.

Grown for ornament and locally naturalized in the Malvern Hills, Kent and Devon. A native of S. Europe and N.W. Africa.

Fruit × 4; fruit t.s. × 4.

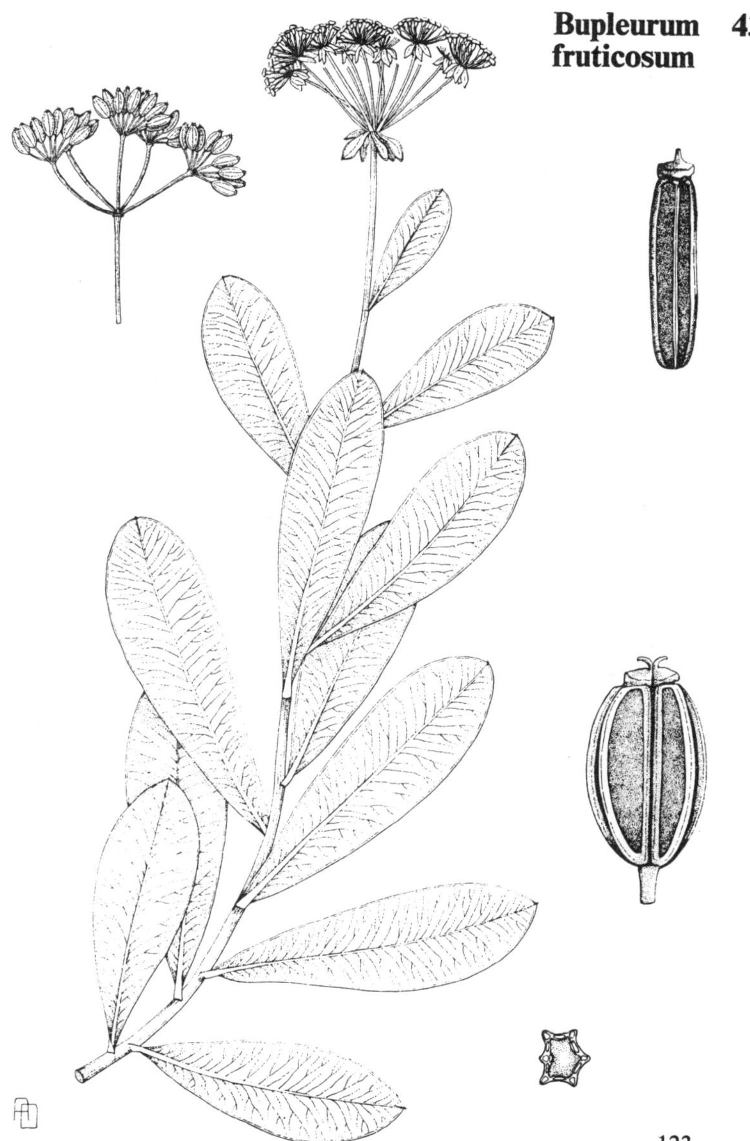

Bupleurum 43
fruticosum

44. **Trinia glauca** (L.) Dumort.

Honewort

A glabrous, glaucous, usually dioecious biennial or perennial, probably always monocarpic. *Stems* up to c. 20 cm, solid, grooved, surrounded by abundant fibrous remains of petioles at the base, much branched, the branches spreading at a wide angle. *Lower leaves* 2- to 3-pinnate, the lobes 5–15 mm, linear, acute, with a cartilaginous apex; petiole slender, sheathing below; *upper leaves* smaller and less divided. *Umbels* compound, the male with usually 4–8 smooth, subequal rays 0.5–1 cm, the female similar but with very unequal rays up to 3 cm; peduncle longer than rays. *Bracts* 0–1, 3-fid; bracteoles 0–several, simple or 2- to 3-fid. *Partial umbels* of male plants dense, with numerous shortly pedicellate flowers, those of female plants lax, with few long-pedicellate flowers. *Flowers* white; sepals absent; outer petals not radiating; styles with enlarged base, forming the stylopodium. *Fruit* c. 3 mm, ovoid, somewhat compressed laterally, smooth; commissure narrow; mericarps with prominent, broad ridges; carpophore present; vittae 5 large within the ridges and 4 small between the ridges; styles 2–3 times as long as the stylopodium, recurved and appressed; stigma capitate. *Cotyledons* tapered gradually at the base, without a distinct petiole. $2n = 18*$. Flowering in May and June.

Dry limestone grassland, very local. S. Devon, N. Somerset and Bristol. C. and S. Europe, northwards to S. England; S.W. Asia.

Represented in Britain by subsp. **glauca,** which occurs throughout the range of the species.

The genus contains about 12 species and extends eastwards to C. Asia.

Fruit × 10; fruit t.s. × 12.

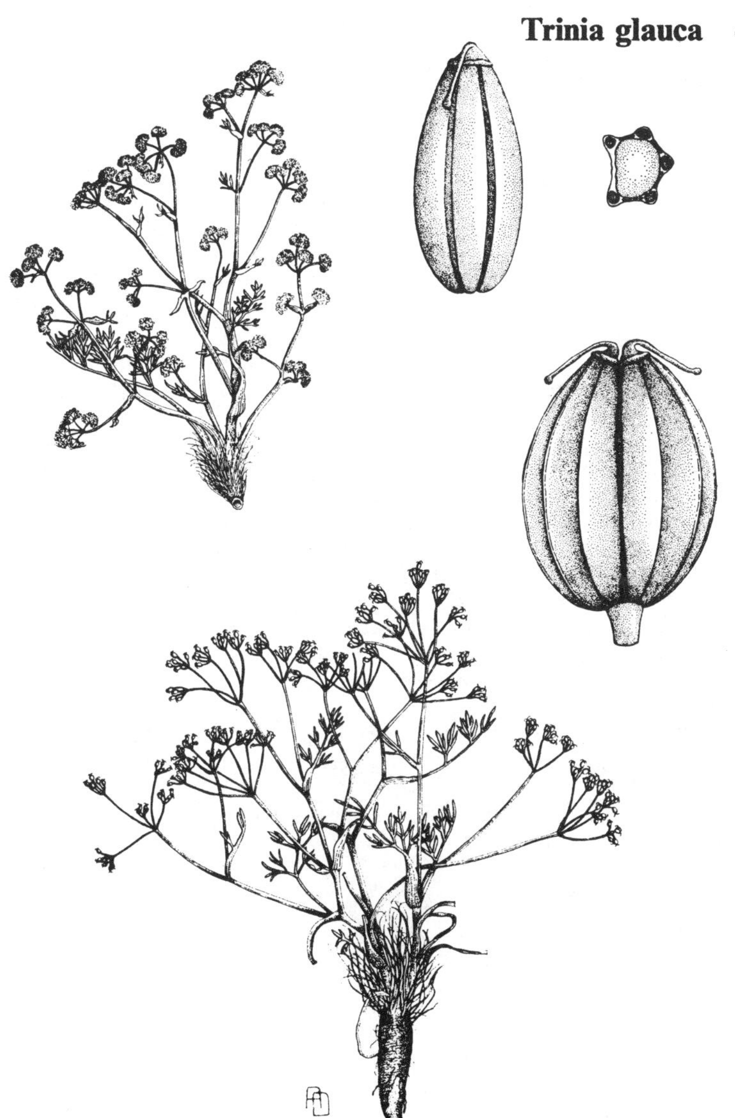

45. **Apium graveolens** L.

Wild Celery

A glabrous biennial with a strong smell of celery. *Stems* up to c. 100 cm, erect, strongly grooved, solid. *Lower leaves* simply pinnate, with 1–3 pairs of stalked, coarsely toothed and sometimes shallowly pinnatifid lobes 1–5 cm, deltate to rhombic in outline; petiole slender, with a sheathing base; *upper leaves* mostly 3-fid or simply pinnate, with 1 pair of lobes and the short petiole sheathing for all or most of its length. *Umbels* compound, with 4–12 smooth subequal rays c. 1(–3) cm long; peduncles shorter than the rays or almost absent, often leaf-opposed; all umbels with hermaphrodite flowers. *Bracts* and bracteoles absent. *Flowers* white; sepals absent; outer petals not radiating; styles with enlarged base, forming the stylopodium. *Fruit* c. 1.5 mm, broadly ovoid, laterally compressed, smooth; commissure narrow; mericarps with prominent slender ridges; carpophore present; vittae solitary; pedicels 2–4 mm; style about as long as the stylopodium, recurved and appressed; stigma a small knob. *Cotyledons* abruptly contracted into a petiole. $2n = 22$. Flowering from June to August.

Coasts of England, Wales and Ireland, local and reaching southern Scotland; in damp often slightly brackish soils, but sometimes inland in the south. Coasts of Europe northwards to c. 56°N; temperate Asia; North Africa. Widely cultivated and often escaping elsewhere.

Var. *dulce* (Miller) DC., celery, is grown for its edible petioles, and var. *rapaceum* (Miller) DC., celeriac, is also cultivated for its swollen edible stock.

The genus contains about 6 species which occur in Europe, temperate Asia and North Africa.

Fruit × 18; fruit t.s. × 18.

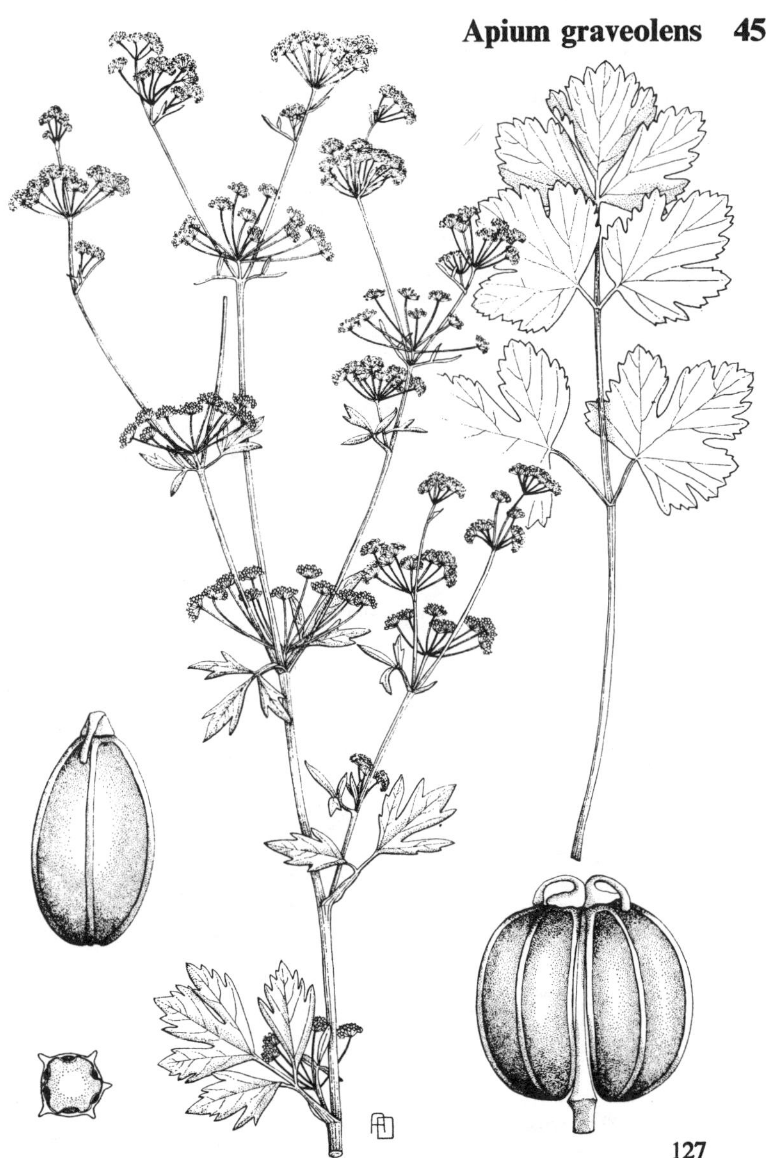

46. **Apium nodiflorum** (L.) Lag.

Fool's Watercress

A glabrous perennial. *Stems* up to 100 cm, procumbent or ascending, rooting at the lower nodes, hollow, finely grooved. *Leaves* simply pinnate, with (1–)2–4(–6) pairs of lobes; lobes (0.3–)1.5–6(–10) cm, lanceolate to ovate, serrate or crenate and often somewhat lobed, sessile. *Umbels* compound, with 3–15 usually subequal, scabrid rays (0.3–)1–2 cm long; peduncles shorter than the rays or almost absent, leaf-opposed; all umbels with hermaphrodite flowers. *Bracts* usually absent, rarely 1–2; bracteoles 4–7, linear-lanceolate to ovate, usually as long as or longer than the flowers. *Flowers* greenish-white; sepals absent; outer petals not radiating; styles with enlarged base, forming the stylopodium. *Fruit* 2–2.5 mm, ovoid, laterally compressed, smooth; commissure narrow; mericarps with prominent thick ribs; carpophore present; vittae solitary; pedicels 1–2 mm; style somewhat longer than the stylopodium, recurved; stigma a small knob. *Cotyledons* abruptly contracted into a petiole. $2n = 22^*$. Flowering in July and August.

In ditches, shallow ponds and other wet places. Common in suitable habitats throughout much of the British Isles, but rare in Scotland and mainly in the south. C. and S. Europe, W. and C. Asia, North Africa.

Rather like *Berula erecta* but easily distinguished by the usual absence of bracts. Sometimes mistaken for watercress, from which it can be distinguished by its taste, hollow stems, inflorescence, flower and fruit.

Fruit × 15; fruit t.s. × 15.

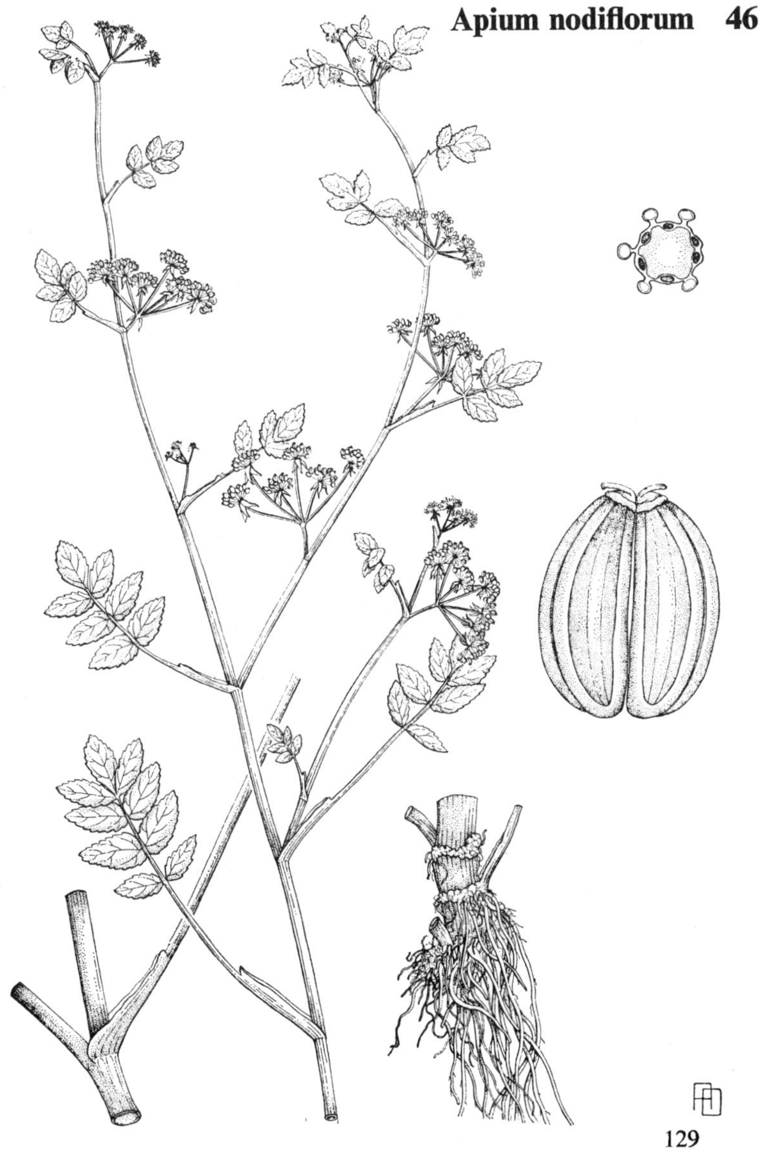

47. **Apium repens**   (Jacq.) Lag.

Creeping Marshwort

A slender perennial with creeping stems rooting at the nodes. *Leaves* simply pinnate, with usually 2–5 pairs of suborbicular, shallowly lobed or toothed, sessile lobes 0.2–1 cm long. *Umbels* compound, with 4–7, usually subequal, smooth rays 0.5–3 cm; peduncles longer than the rays, leaf-opposed; all umbels with hermaphrodite flowers. *Bracts* 3–7, lanceolate to ovate, deflexed; bracteoles like the bracts, usually shorter than the flowers. *Flowers* white; sepals absent; outer petals not radiating; styles with enlarged base, forming the stylopodium. *Fruit* c. 1 mm, suborbicular, slightly broader than long, laterally compressed, smooth; commissure narrow; mericarps with prominent slender ribs; carpophore present; vittae solitary; pedicels 1–2 mm; style about twice as long as the stylopodium, recurved; stigma a small knob. $2n = 16^*$. Flowering in July.

Old damp meadows, ditches and shallow ponds, very local. Known from 4 localities in Oxfordshire. The general distribution is somewhat uncertain owing to confusion with small variants of *A. nodiflorum*, but it appears to occur mainly in C. Europe.

The presence of 3–7 bracts, the longer peduncle and the smaller fruit distinguish *A. repens* from *A. nodiflorum*, with which it sometimes hybridizes.

Fruit $\times$ 30; fruit t.s. $\times$ 30.

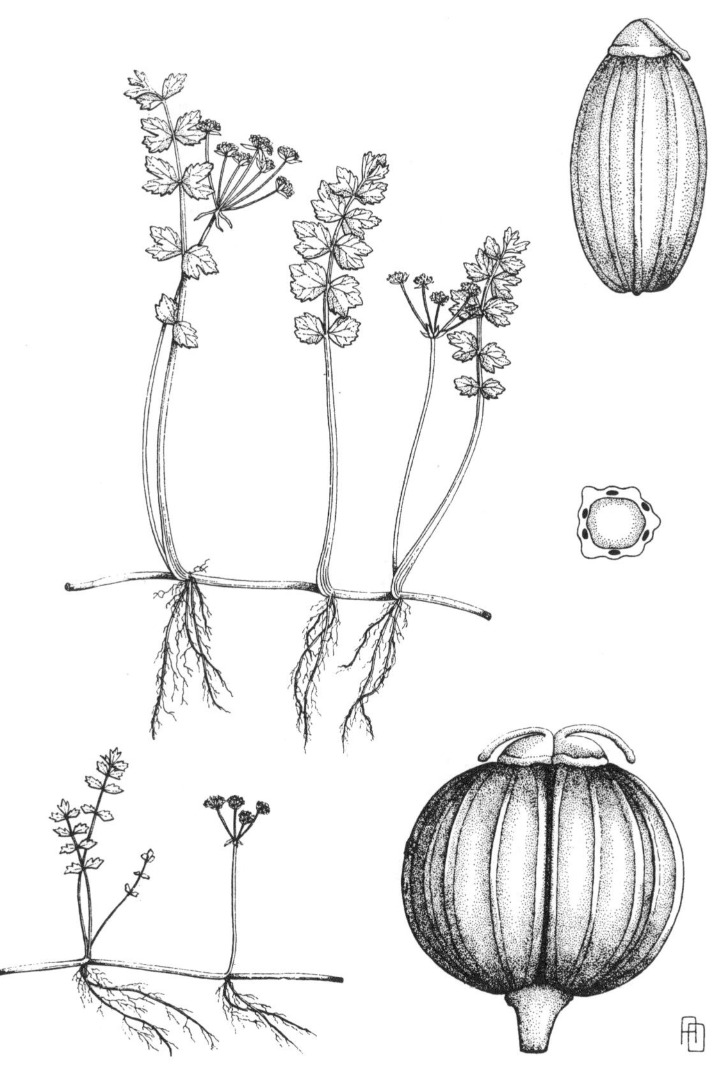

48. **Apium inundatum** (L.) Reichenb. fil.

Lesser Marshwort

A usually partly or completely submerged glabrous perennial. *Stems* up to 75 cm, slender, weak, smooth. *Submerged leaves*, and lower leaves of terrestrial plants, 2- to 3-pinnate, with linear lobes, the upper pinnate, with lanceolate to ovate, often 3-lobed sessile lobes c. 5 mm long. *Umbels* compound, with 2(–4) smooth rays usually 0.5–1 cm; peduncles short, leaf-opposed: all umbels with hermaphrodite flowers. *Bracts* absent; bracteoles 3–6, lanceolate, not concealing the flowers and fruit. *Flowers* white; sepals absent; outer petals not radiating; styles with enlarged base, forming the stylopodium. *Fruit* 2.5–3 mm, elliptic-oblong, laterally compressed, smooth; commissure narrow; mericarps with prominent ridges; carpophore present; vittae solitary; pedicels very short; styles about half as long as the stylopodium, spreading; stigma a small knob. $2n = 22$. Flowering from June to August.

In lakes, ponds and ditches. Local but widely distributed in the British Isles. W. Europe, extending eastwards to Sicily and S.E. Sweden.

The hybrid between *A. inundatum* and *A. nodiflorum* (**A. × moorei** (Syme) Druce) occurs fairly frequently in Ireland, rarely in England, and is apparently unknown elsewhere. It usually has all leaves, whether submerged or not, simply pinnate, usually with obovate lobes, and the umbels usually have 2–3 rays. It is sterile and flowers rather sparingly.

Fruit × 10; fruit t.s. × 15.

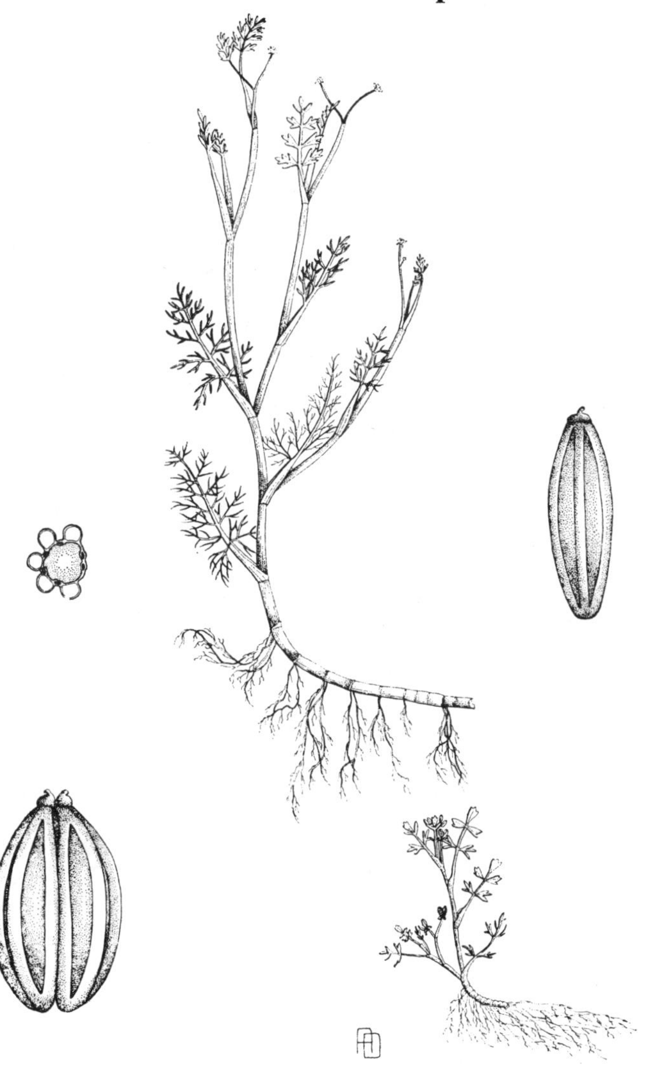

49. **Trachyspermum ammi** (L.) Sprague

An almost glabrous annual. *Stems* usually 10–30 cm, hollow, striate, much branched. *Lower leaves* 2- to 3-pinnatisect, withered by flowering time, long-petiolate, the lobes 1–2 cm, linear or filiform; *upper leaves* similar but smaller and sometimes simply pinnatisect, shortly petiolate, with a short, sheathing base. *Umbels* compound, with 5–10 (–20) smooth or sparsely puberulent rays c. 1 cm; peduncles longer than the rays; all umbels with hermaphrodite flowers. *Bracts* usually 4–5, linear, sometimes lobed; bracteoles 3–6, linear-lanceolate, sometimes sparsely puberulent. *Flowers* white; sepals small, acute; petals hairy beneath, the outer not radiating; styles with enlarged base, forming the stylopodium. *Fruit* 1.5–2 mm, ovoid, laterally compressed, irregularly covered with grey papillae; commissure narrow; mericarps with prominent ridges; carpophore present; vittae solitary; pedicels 1–4 mm, hairy; styles about twice as long as the stylopodium, recurved; stigma capitate. $2n = 18$. Flowering in September.

Grown in east Mediterranean countries for its aromatic fruit, which is used as a flavouring. An occasional casual on rubbish tips in England. It is apparently not known wild.

The densely grey-papillose ovary and fruit combined with the dissected leaves and white petals distinguish it from all other British umbellifers.

The genus contains about 20 species, which occur in Africa and Asia.

Fruit × 15; fruit t.s. × 15.

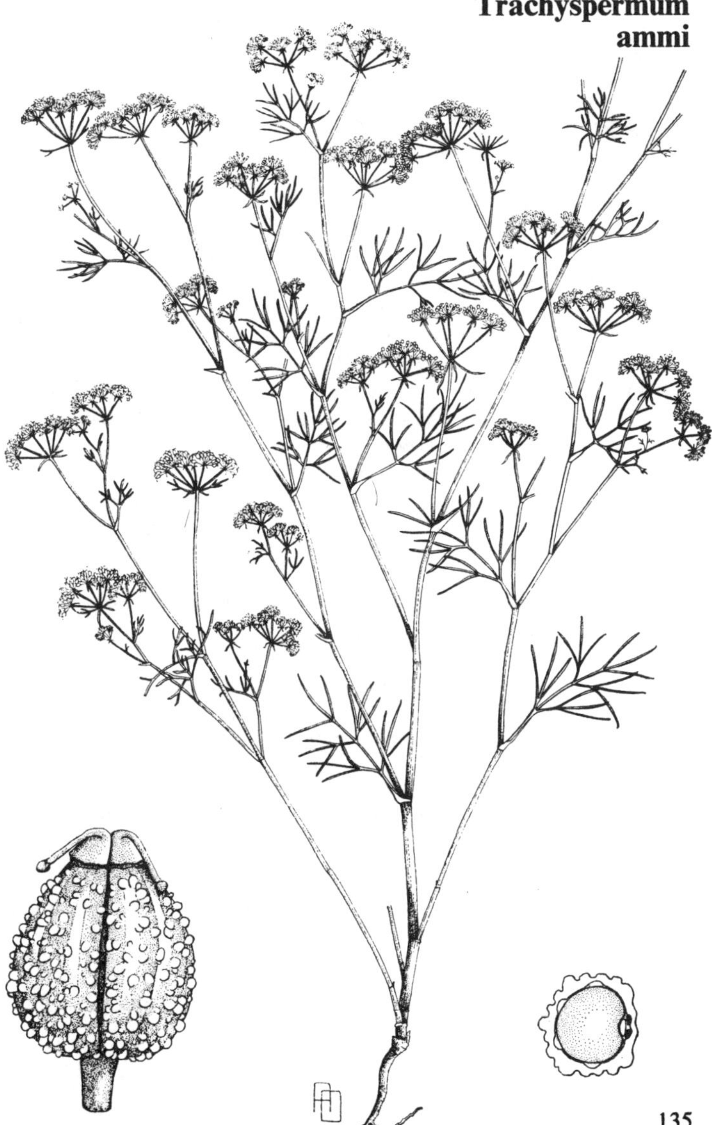

50. **Petroselinum crispum**   (Miller) A. W. Hill

Garden Parsley

A glabrous biennial. *Stems* up to c. 75 cm, erect, striate, solid, often with some sheathing remains of petioles at the base. *Lower leaves* triangular in outline, 3-pinnate, shiny, the lobes 10–30 mm, ovate in outline, usually cuneate at the base, toothed and often more or less 3-fid, strongly crispate in cultivars; petiole long, sheathing near the base; *upper cauline leaves* small, 1- to 2-pinnate, with short, mostly or entirely sheathing and usually broadly scarious petioles. *Umbels* compound, with 8–20 smooth, subequal rays 0.5–3 cm; peduncle longer than the rays; terminal umbel with hermaphrodite flowers, the lateral umbels with male and hermaphrodite flowers, the latest to develop with almost entirely male flowers. *Bracts* 1–3, entire or 2- to 3-fid and more or less leaf-like; bracteoles 5–8, linear-oblong to ovate-cuspidate, usually with a broad scarious margin. *Flowers* yellowish; sepals absent; outer petals not radiating; styles with enlarged base, forming the stylopodium. *Fruit* 2–2.5 mm, ovoid to ellipsoid, laterally compressed, smooth; commissure narrow; mericarps with rather slender low ridges; carpophore present; vittae solitary, conspicuous; pedicels 2–4 mm, rather stout; styles about as long as the stylopodium, recurved and appressed; stigma a small knob. *Cotyledons* abruptly contracted into a petiole. $2n = 22*$. Flowering from June to August.

Walls, rocks and waste places; escaped from cultivation and naturalized in much of Europe. Origin uncertain.

Cultivars have more or less crisped (curled) leaves and naturalized plants often show this character.

Petroselinum contains the two species included here and possibly a third species, and is native in Europe, W. Asia and North Africa.

Fruit × 12; fruit t.s. × 15.

51. **Petroselinum segetum** (L.) Koch

Corn Parsley

A glabrous, more or less glaucous biennial or annual. *Stems* up to 100 cm, striate, solid. *Leaves* simply pinnate, mostly near the base of the plant, linear-oblong in outline, with 4–12 pairs of suborbicular to lanceolate, serrate or sometimes shallowly lobed, sessile lobes 0.5–3.5 cm, the obtuse teeth with a cartilaginous margin and mucro c. 0.5 mm long; petiole slender, not dilated much at the base. *Umbels* compound, with 3–10 smooth, very unequal rays up to c. 3 cm; peduncle longer than the rays; all umbels with hermaphrodite flowers. *Bracts* and bracteoles 2–5, subulate. *Flowers* white; sepals small; outer petals not radiating; styles with enlarged base, forming the stylopodium. *Fruit* 2.5–3 mm, ovoid, laterally compressed, smooth; commissure narrow; mericarps with prominent, stout ridges; carpophore present; vittae solitary, conspicuous; some pedicels very short, others up to 15 mm long; styles much shorter than the stylopodium, divergent; stigma a small knob. $2n = 18$. Flowering in August and September.

Arable fields, hedge-banks, river-banks and brackish grassland. S. and E. England, north to the Humber; a few coastal localities in Wales. W. Europe, from the Netherlands to Portugal and eastwards to C. Italy, rare and apparently decreasing.

The stem and fruit smell like parsley when crushed.

Fruit × 10; fruit t.s. × 10.

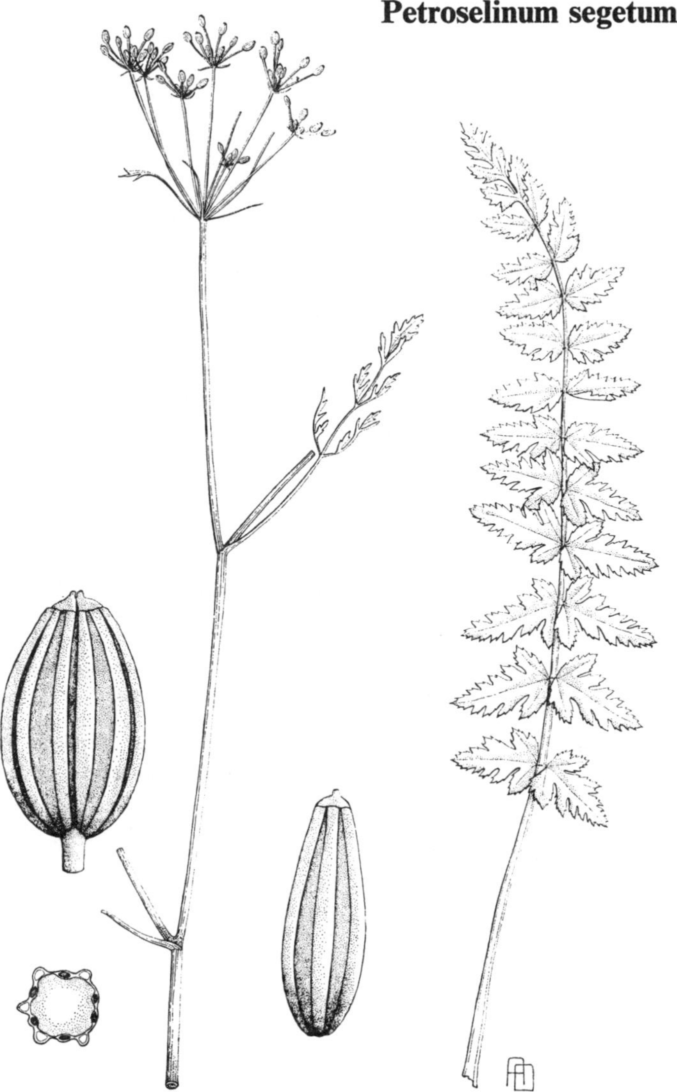

52. **Sison amomum** L.

Stone Parsley

A glabrous biennial with a nauseous smell when crushed. *Stems* up
to c. 100 cm, solid, striate. *Leaves* simply pinnate, with the lowest
lobes sometimes pinnatifid or rarely pinnatisect; lobes of lower
leaves 2–5 pairs, 3–6 cm, lanceolate to ovate-lanceolate, cuneate at
the base, sessile or subsessile, serrate, the obtuse teeth with a cartilaginous margin and short, acute mucro; petiole long, with a
gradually dilated base; upper cauline leaves mostly with 1–2 pairs of
linear to narrowly spathulate, toothed lobes and a short, sheathing
petiole. *Umbels* compound, with 3–6 smooth rays up to 3 cm, one of
which is usually much shorter than the remainder; peduncle longer
than the rays; all umbels with hermaphrodite flowers. *Bracts* 2–4,
linear-lanceolate to subulate; bracteoles 2–4, usually ovate-lanceolate. *Flowers* white; sepals absent; outer petals not radiating; styles
with enlarged base, forming the stylopodium. *Fruit* c. 3 mm, subglobose, somewhat compressed laterally, smooth; commissure
narrow; mericaps with fairly prominent ridges; carpophore present;
vittae c. $\frac{1}{2}$ as long as the mericarp, solitary, conspicuous; pedicels
1–5 mm, somewhat unequal; styles c. $\frac{1}{2}$ as long as the stylopodium,
divergent to recurved; stigma a small knob. *Cotyledons* abruptly
contracted into a petiole. $2n = 14$. Flowering from July to September.

S. and E. England, north to the Humber; a few coastal localities
in Wales. W. and S. Europe, S.W. Asia, North Africa.

Remarkably like *Petroselinum segetum*, from which it is most
easily distinguished by the smell, somewhat like that of petrol, the
larger lobes of the basal leaves, the finely divided upper cauline
leaves and the umbels with less unequal rays.

Sison contains 2 species, this and *S. exaltatum* Boiss. from the
eastern Mediterranean region.

Fruit × 15; fruit t.s. × 10.

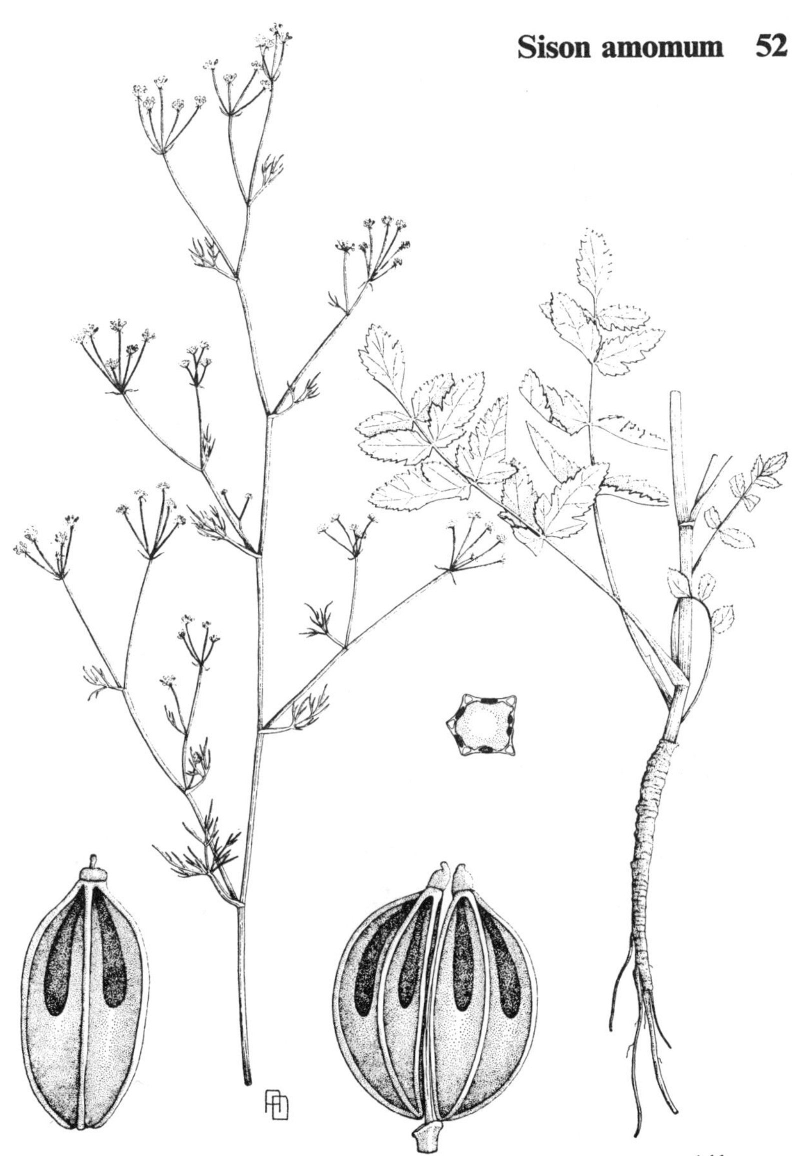

53. **Cicuta virosa** L.

Cowbane

A glabrous perennial with an ovoid or shortly cylindrical septate stock. *Stems* up to c. 150 cm, hollow, striate. *Leaves* 2- to 3-pinnate, the lobes 3–9 cm, linear-lanceolate, unequal at the base, rather remotely and deeply serrate; midrib scabrid above, smooth or almost so beneath; petiole long, hollow, sheathing but not conspicuously dilated at the base; upper cauline leaves similar to the lower, but smaller. *Umbels* compound, with 10–30 subequal smooth rays usually 2–6 cm long; peduncle somewhat longer than the rays; all umbels with mostly hermaphrodite flowers. *Bracts* absent; bracteoles numerous, linear-oblong. *Flowers* white; sepals ovate, conspicuous; outer petals not radiating; styles with enlarged base, forming the stylopodium. *Fruit* 1.75–2 mm, subglobose, somewhat wider than long, subterete, smooth; commissure wide; mericarps with low, very wide ridges; carpophore present; vittae slender, solitary, conspicuous; pedicels c. 6 mm, slender; styles 2–3 times as long as the stylopodium, curved; stigma a small knob. *Cotyledons* tapered gradually at the base, without a distinct petiole. $2n = 22$. Flowering in July and August.

In shallow water in ponds, ditches etc., local. Mainly in East Anglia, Shropshire, Cheshire and S.E. Scotland; in Ireland in the centre and north of the Republic. Europe, mainly north of 45°N, temperate Asia.

Very poisonous due to cicutoxin, a convulsant poison. It has occasionally been eaten in mistake for parsnips, with fatal results, though cattle are more frequently poisoned.

The only other aquatic umbellifer with which *Cicuta* might be confused is *Sium latifolium*, which has simply pinnate leaves and the umbel with bracts.

There are about 10 species of *Cicuta* in north temperate regions.

Fruit × 15; fruit t.s. × 7.

54. **Ammi majus** L.

Bullwort

A glabrous annual. *Stems* up to c. 100 cm, solid, striate. *Leaves* very variable, the lower 1– to 2-pinnate,with ovate to lanceolate, serrate, obtuse lobes, the upper usually 2-pinnate, with lanceolate to linear, often pinnatifid or deeply toothed acute lobes; petiole gradually widened to the sheathing base. *Umbels* compound, with 9–40 slender, somewhat scabrid, subequal rays 3–7 cm; peduncle longer than the rays; all umbels with hermaphrodite flowers. *Bracts* several, $\frac{1}{3}$–$\frac{3}{4}$ as long as the rays, usually leaf-like, 3-fid to pinnatisect, with linear lobes, occasionally some entire; bracteoles numerous, usually lanceolate, with a long filiform apex and broad scarious margin. *Flowers* white; sepals absent; outer petals slightly radiating; styles with enlarged base, forming the stylopodium. *Fruit* 1.5–2 mm, ellipsoid, slightly compressed laterally, smooth; commissure narrow; mericarps with slender, prominent ridges; carpophore present; vittae solitary, conspicuous; pedicels 1–6 mm, slightly scabrid; styles about twice as long as the stylopodium, recurved; stigma a small knob. *Cotyledons* tapered at the base, without a distinct petiole. $2n = 22$. Flowering from June to October.

A rather uncommon casual on rubbish tips, in waste places and as a wool alien. Native of S. Europe, North Africa and Asia Minor.

A. visnaga (L.) Lam. occurs occasionally in waste places. It is like *A, majus* but the rays thicken and become erect after flowering, while the bracts are deflexed. It has $2n = 20$, flowers from August to October, and is native in the Mediterranean region.

The genus contains about 10 species in the Mediterranean region and the Atlantic Islands.

Fruit (left) × 15, (right) × 17; fruit t.s. × 15.

55. **Falcaria vulgaris** Bernh.

Longleaf

A glaucous freely branched perennial. *Stems* up to c. 90 cm, solid, striate. *Leaves* 3-fid, one or more of the lobes often again 2- or 3-fid; lobes up to c. 30 cm, linear-lanceolate to linear, often somewhat falcate, acute, regularly and sharply serrate, often puberulent beneath, with a thick cartilaginous margin, the midrib usually with another vein parallel with and close to it on each side; petiole slender, gradually widened to the sheathing base. *Umbels* compound, with 9–18 slender, smooth, subequal rays 2–4 cm long; peduncle usually longer than the rays, often leaf-opposed; terminal umbels with hermaphrodite flowers, the lateral with male flowers. *Bracts* and bracteoles 4–15, subulate. *Flowers* whitish; sepals conspicuous; outer petals not radiating; styles with enlarged base, forming the stylopodium. *Fruit* 3–4 mm, oblong, laterally compressed, smooth; commissure narrow; mericarps with low ridges, wider than the grooves; carpophore present; vittae solitary, slender; pedicels 4–5 mm, slender; styles about twice as long as the stylopodium, recurved; stigma a small knob. *Cotyledons* abruptly contracted into a petiole. $2n = 22$. Flowering from July to September.

Locally naturalized, mainly in S. England and the Channel Islands. Native in Europe from N. France and C. Russia southward, and in W. Asia.

Sometimes annual or biennial, but in this country probably always perennial. The long, narrow, regularly and sharply serrate leaf-lobes are characteristic.

The genus contains about 3 species in Europe and Asia.

56. **Carum carvi** L.

Caraway

A glabrous biennial. *Stems* usually 30–60 cm, hollow, striate. *Leaves* 2- to 3-pinnate, the lobes usually 5–10 mm, often pinnatifid, linear-lanceolate to linear, acute, with a cartilaginous apex; petiole rather short, with a broad, scarious sheathing base. *Umbels* compound, with 5–16 smooth, markedly unequal rays 1–6 cm long; peduncle shorter to longer than the rays, often leaf-opposed; umbels with mostly hermaphrodite flowers. *Bracts* and bracteoles absent or few, the former occasionally leaf-like. *Flowers* usually whitish, sometimes pink or red; sepals absent; outer petals slightly radiating; styles with enlarged base, forming the stylopodium. *Fruit* 3–4 mm, ellipsoid, laterally compressed, smooth, with a distinctive smell when crushed; commissure narrow; mericarps with narrow, low ridges; carpophore present; vittae solitary, wide; pedicels 5–15 mm; styles a little longer than the stylopodium, recurved and appressed; stigma capitate. *Cotyledons* tapered gradually at the base, without a distinct petiole. $2n = 20$. Flowering in June and July.

Widely distributed but rather uncommon in the British Isles. Cultivated for the fruits, which are used for flavouring, and naturalized in waste places; perhaps native in some localities. *C. carvi* occurs throughout most of Europe, temperate Asia and N.W. Africa, but is rare in the Mediterranean region. It is naturalized in most temperate countries.

The genus occurs in Europe, temperate Asia and North Africa and contains about 30 species; this species can be easily recognized by the smell of its fruits when crushed.

Fruit × 10; fruit t.s. × 10.

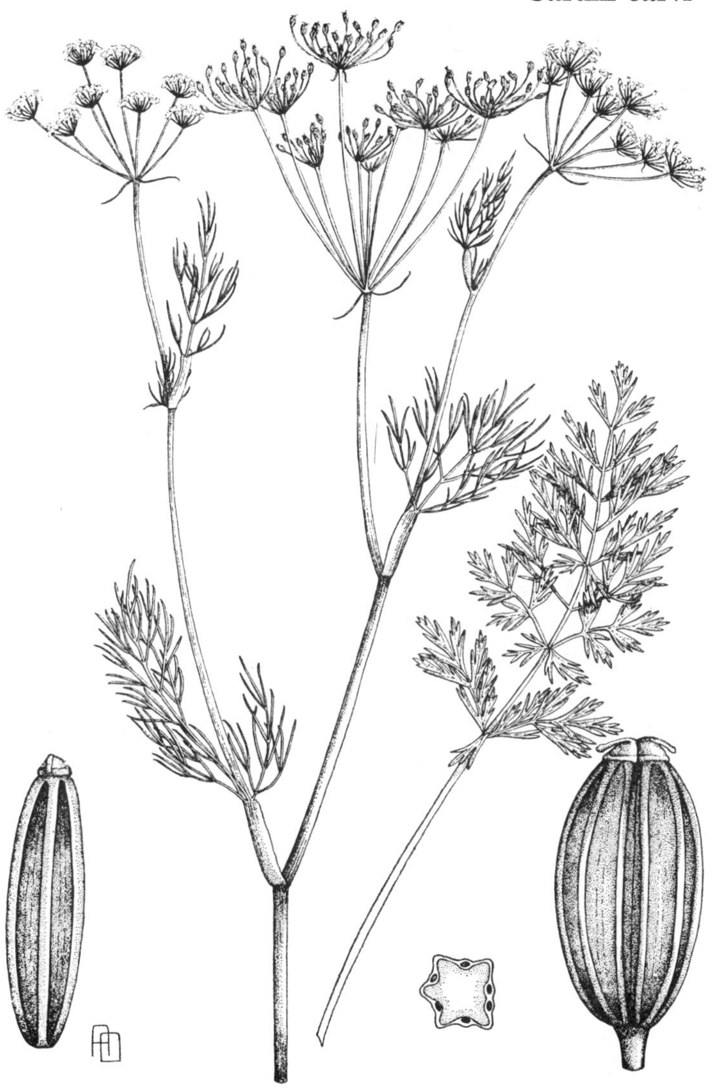

57. **Carum verticillatum** (L.) Koch

Whorled Caraway

A glabrous perennial with fusiform roots thickened downwards. *Stems* up to c. 60 cm, hollow, striate, surrounded at the base by fibrous remains of petioles. *Leaves* narrowly oblong in outline, mostly basal, simply pinnate, with usually more than 20 pairs of deeply palmatisect segments; lobes filiform, appearing as if whorled, the lowest c. 1 mm, the upper up to c. 10 mm; petiole short, scarcely enlarged at the base. *Umbels* compound, with usually 8–14 smooth, subequal rays 1.5–4 cm long; peduncle longer than the rays, leaf-opposed; umbels with mostly hermaphrodite flowers. *Bracts* up to 10, linear to lanceolate, acute, deflexed; bracteoles like the bracts, but not deflexed. *Flowers* white; sepals small; outer petals not radiating; styles with enlarged base, forming the stylopodium. *Fruit* 2.5–3 mm, ellipsoid, laterally compressed, smooth; commissure narrow; mericarps with conspicuous narrow ridges; carpophore present; vittae solitary, wide; pedicels 3–5 mm; styles a little longer than the stylopodium, recurved; stigma slightly thickened. *Cotyledons* tapered gradually at the base, without a distinct petiole. $2n = 20$. Flowering in July and August.

Marshes, streams and damp meadows, calcifuge. Very local in the west of England, Wales and Scotland and N. and W. Ireland; one locality in Surrey and one in Aberdeenshire. W. Europe, northwards to Scotland and the Netherlands.

The narrowly oblong leaves with usually more than 20 pairs of deeply palmatisect segments are characteristic of *C. verticillatum*.

Fruit × 12; fruit t.s. × 10.

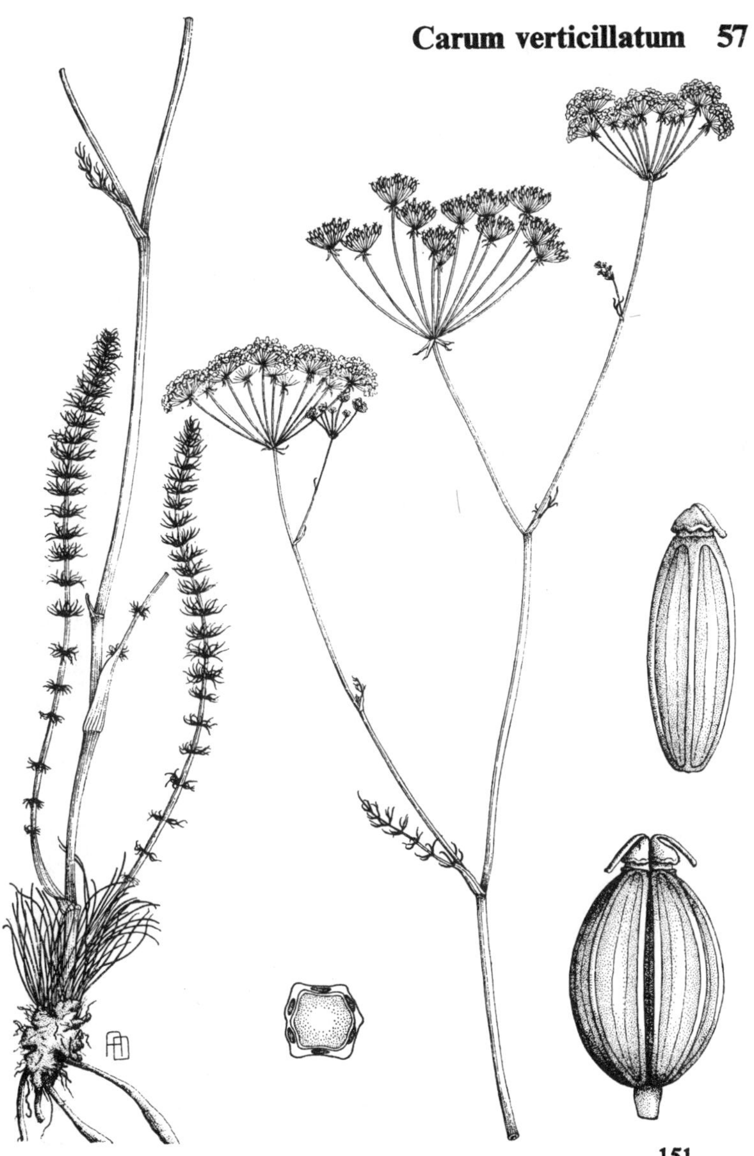

151

58. **Selinum carvifolia** (L.) L.

Cambridge Milk-Parsley

A glabrous perennial. *Stems* up to c. 100 cm, solid, grooved and strongly angled, the angles narrowly winged. *Leaves* 2- to 3-pinnate, lanceolate in outline, the lobes 3–10 mm, linear-lanceolate to ovate, minutely serrulate, sometimes lobed, with an acute cartilaginous apex; petiole long, somewhat dilated at the base. *Umbels* compound, with usually 15–25 somewhat unequal rays 1.5–4 cm, papillose on the angles; peduncle longer than the rays, papillose on the angles near the top; umbels with mostly hermaphrodite flowers. *Bracts* usually absent, sometimes few and soon falling; bracteoles c. 10, subulate to linear-lanceolate. *Flowers* white; sepals minute; outer petals not radiating; styles with enlarged base, forming the stylopodium. *Fruit* 3–4 mm, ovoid-oblong, dorsally compressed, smooth; commissure narrow; mericarps with the 3 dorsal ridges narrowly winged and the lateral broadly winged; carpophore present; vittae solitary; pedicels as long as to twice as long as the bracteoles, papillose; styles about twice as long as the stylopodium, recurved and usually appressed; stigma capitate. *Cotyledons* tapering gradually at the base, without a distinct petiole. $2n = 22$. Flowering from July to October.

Fens and damp meadows, very local. Cambridgeshire; formerly in Nottinghamshire and N. Lincolnshire, but now apparently extinct there. Most of Europe, except for much of the Mediterranean region, eastwards to C. Asia.

There are about 4 species of *Selinum* in Europe and Asia.

59. **Ligusticum scoticum** L.

Scots Lovage

A glabrous, bright green perennial. *Stems* up to c. 90 cm, hollow, striate, often reddish towards the base. *Leaves* 1- to 2-ternate, the primary divisions long-stalked, the lobes usually sessile; lobes 2–5 cm, rhombic to ovate, coarsely toothed or shallowly lobed in the upper half, entire in the lower half, the teeth and apex obtuse; petiole of lower leaves long, with a broadly sheathing base, of upper leaves entirely sheathing. *Umbels* compound, with 8–14 subequal rays 1.5–4 cm long, papillose near the top; peduncle longer than the rays, papillose near the top; umbels with mostly hermaphrodite flowers. *Bracts* 1–5, linear, membranous; bracteoles usually c. 7, linear to linear-lanceolate. *Flowers* greenish-white, sometimes tinged with pink; sepals triangular; outer petals not radiating; styles with enlarged base, forming the stylopodium. *Fruit* 4–7 mm, oblong-ovoid, not compressed, smooth; commissure broad; mericarps with prominent narrowly winged ridges; carpophore present; vittae several; pedicels about as long as the bracteoles, papillose; styles about as long as the stylopodium, recurved; stigma truncate. *Cotyledons* tapered gradually at the base, without a distinct petiole. $2n = 22$. Flowering in June and July.

Rocky coasts, local. Scotland, northern half of Ireland; perhaps extinct in Northumberland; very susceptible to grazing by sheep. Coasts of N. Europe, Greenland and eastern North America.

The ternate leaves with glabrous lobes 2–5 cm, rhombic in outline and often 3-fid at the apex, in conjunction with white flowers distinguish this from other British umbellifers.

According to J. D. Hooker the leaves were formerly eaten as a pot herb and the root is aromatic and pungent.

The genus contains about 30 species in north temperate regions.

Fruit × 5; fruit t.s. × 7.

60. **Angelica sylvestris** L.

Wild Angelica

A nearly glabrous perennial. *Stems* up to 200 cm or more, hollow, often purplish and pruinose, striate. *Leaves* 2- to 3-pinnate, the primary divisions long-stalked, the lobes usually sessile; lobes 1.5–8 cm, obliquely ovate to lanceolate, usually acute or cuspidate, serrate or biserrate, with cartilaginous teeth, usually more or less hispidulous on both surfaces at least on the veins; lower leaves with long, laterally compressed petioles deeply channelled on the upper side and inflated at the base; upper leaves with a strongly inflated petiole and the lamina very small or absent. *Umbels* compound, with 15–40 subequal rays usually 2–8 cm, the rays and peduncle densely puberulent on the ridges and sparsely so between them; umbels with hermaphrodite flowers. *Bracts* absent or few and caducous; bracteoles usually 6–10, linear, puberulent. *Flowers* white or pinkish. Sepals minute; outer petals not radiating; styles with enlarged base, forming the stylopodium. *Fruit* 4–5 mm, ovate, dorsally compressed, smooth; commissure narrow; mericarps with prominent, obtuse dorsal ridges and broadly winged lateral ridges, the wings being wider than the mericarp, undulate and not closely appressed to one another; carpophore present; vittae solitary; pedicels 5–10 mm, puberulent; styles about 3 times as long as the stylopodium, recurved; stigma capitate. *Cotyledons* abruptly contracted into a petiole. $2n = 22$. Flowering from July to September.

Damp grassy places, fens, open woods, by streams and ditches; common and generally distributed. Almost throughout Europe and temperate Asia.

The usually purplish, pruinose, hollow stem and densely puberulent peduncle distinguish this from other umbellifers with similar large-leaf lobes. The hemispherical umbels are also distinctive among roadside umbellifers.

A variant which is about 10 cm high when in flower occurs in turloughs in W. Ireland; whether this is a genetic dwarf does not appear to be known. Somewhat similar variants occur in very exposed positions in W. Scotland.

The genus contains probably about 60 species occurring in north temperate regions and New Zealand.

Fruit × 7; fruit t.s. × 7.

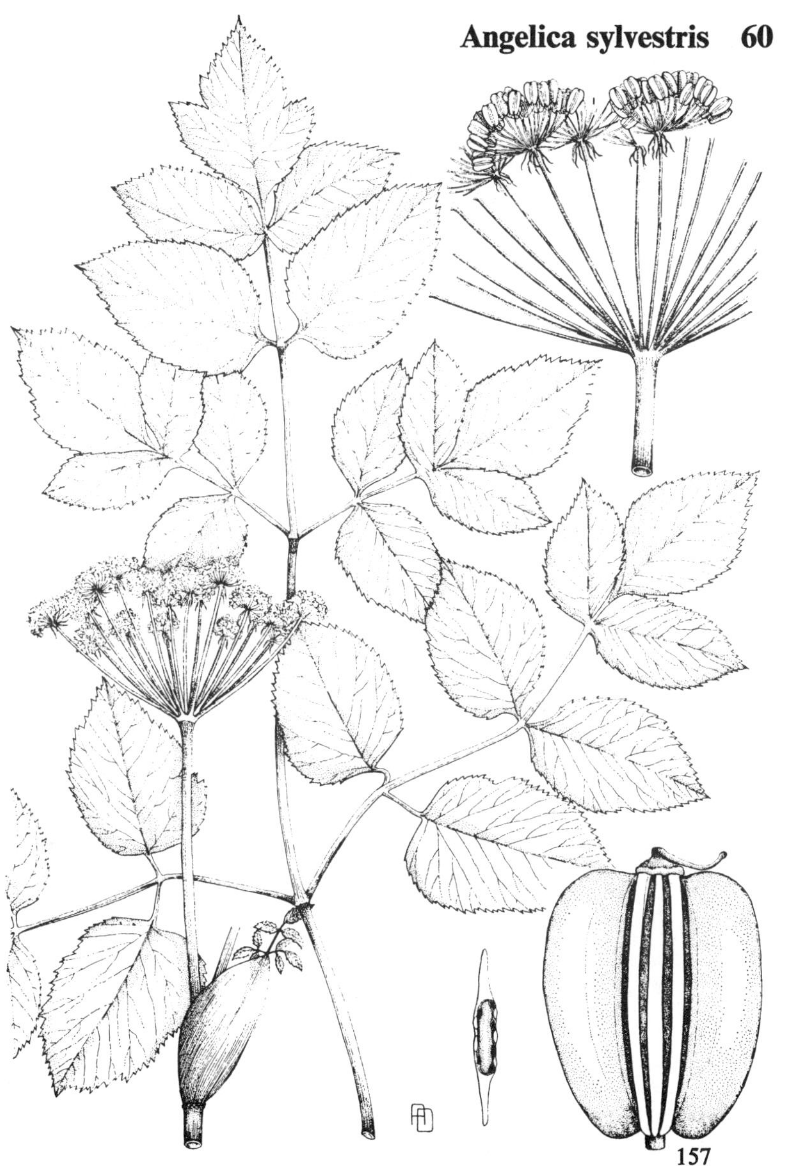

Angelica sylvestris 60
157

61. **Angelica archangelica** L.

Garden Angelica

A robust perennial. *Stems* up to 200 cm or more, hollow, usually green, striate. *Leaves* 2- to 3-pinnate, the primary divisions with rather short stalks, the lobes usually sessile and often decurrent; lobes up to c. 15 cm, deeply toothed and often irregularly lobed, the terminal frequently 3-fid, the margin and apex of the teeth cartilaginous, usually glabrous on both surfaces; lower leaves with a long petiole, inflated at the base; upper leaves with a strongly inflated petiole and the lamina very small or absent. *Umbels* compound, with usually c. 40 somewhat puberulent subequal rays 4–8 cm long; peduncle longer than the rays, glabrous; umbels with hermaphrodite flowers. *Bracts* absent or few and caducous; bracteoles numerous, linear. *Flowers* greenish; sepals small; outer petals not radiating; styles with enlarged base, forming the stylopodium. *Fruit* c. 6 mm, ovate-oblong, dorsally compressed, smooth; commissure narrow; mericarps with prominent, acute dorsal ridges and winged lateral ridges, the wings narrower than the mericarp, rather thick and corky and not closely appressed to one another; carpophore present; vittae solitary; pedicels 10–15 mm, minutely papillose; styles about twice as long as the stylopodium, recurved; stigma capitate. $2n = 22$. Flowering in May and June.

Naturalized on river-banks and in waste places, locally but often abundantly; cultivated for its aromatic petioles used in confectionary and for a liqueur. Native of N. and E. Europe eastwards to C. Asia; Greenland.

Represented in Britain by subsp. **archangelica,** which occurs almost throughout the range of the species.

Very similar to *A. sylvestris*, but the stems are usually green, the peduncle glabrous and the flowers greenish-white.

Fruit × 7; fruit t.s. × 4.

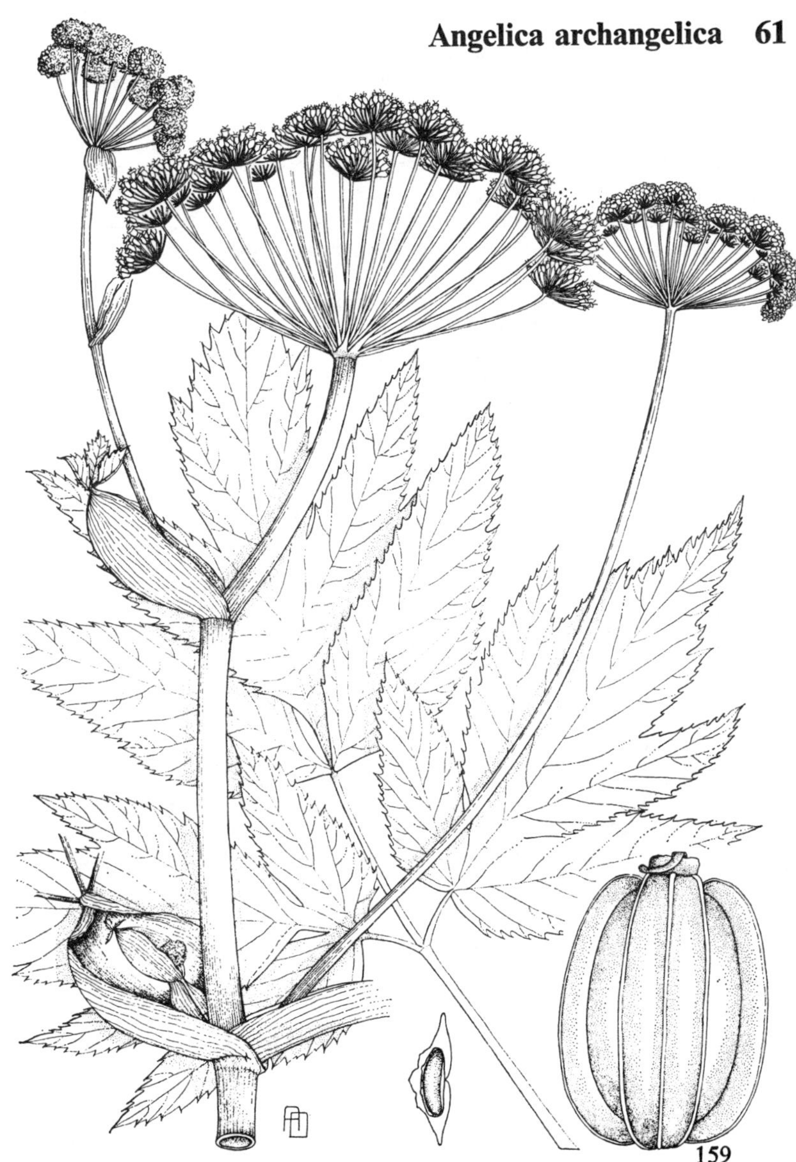

159

62. **Levisticum officinale** Koch

Lovage

A robust, almost glabrous perennial smelling strongly of celery. *Stems* up to c. 200 cm, hollow, striate, surrounded at the base by numerous scale-like remains of petioles. *Leaves* 2- to 3-pinnate, the lobes up to c. 11 cm, rhombic in outline, the cuneate basal half entire, the upper half coarsely incise-serrate or almost lobed, the teeth with a cartilaginous apex; petiole hollow, slightly inflated near the base. *Umbels* compound, with 12–20 stout, subequal rays which are papillose-puberulent on the adaxial surface; peduncles usually longer than the rays, those of the upper lateral umbels often opposite or in whorls of 3–4; umbels with hermaphrodite flowers. *Bracts* numerous, linear-lanceolate, long-acute, deflexed, with a scarious margin; bracteoles similar in shape, not deflexed, almost entirely scarious, connate at least near the base. *Flowers* yellowish; sepals minute; outer petals not radiating; styles with enlarged base, forming the stylopodium. *Fruit* 5–7 mm, broadly elliptical, dorsally compressed, smooth; commissure broad; mericarps with prominent dorsal ridges and winged lateral ridges, the wings narrower than the mericarp; carpophore present; vittae solitary; pedicels usually shorter than the fruit; styles recurved; stigma truncate. $2n = 22$. Flowering in July and August.

Naturalized in a few localities; sometimes cultivated for flavouring. Naturalized in much of Europe, mainly in mountain regions. Probably native in Iran.

Somewhat resembling *Angelica* but easily distinguished by the leaves whose lobes have cuneate, entire proximal halves, the yellowish flowers and the truncate, not capitate stigmas.

Fruit × 7; fruit t.s. × 7.

63. **Peucedanum officinale**   L.

Hog's Fennel, Sulphur-Weed

A glabrous perennial with a stout stock. *Stems* up to c. 200 cm, solid, striate, sometimes weakly angled, surrounded by fibrous remains of petioles at the base. *Lower leaves* 3 to 6 times ternately divided, the primary divisions long-stalked; lobes 4–10 cm, linear, sessile, attenuate at both ends, not all lying in the same plane, with a narrow cartilaginous margin, usually serrulate near the apex; petiole long, not much inflated at the base; *upper leaves* less divided and smaller, with an oblong, sheathing petiole. *Umbels* compound, with usually 15–45 smooth, rather unequal rays 1.5–8 cm long; peduncle longer than the rays, glabrous; umbels with hermaphrodite flowers. *Bracts* absent or few, linear and usually caducous; bracteoles several, linear. *Flowers* yellow; sepals conspicuous, acute; outer petals not radiating; styles with enlarged base, forming the stylopodium. *Fruit* c. 7 mm, elliptical to obovate dorsally compressed, smooth; commissure broad; mericarps with rather prominent dorsal ridges and winged lateral ridges, the wings flat and closely appressed; carpophore present; vittae solitary; at least some pedicels several times as long as the fruit; styles about as long as the stylopodium, stout, recurved; stigma spathulate. *Cotyledons* abruptly contracted into a petiole. $2n = 66$. Flowering from July to September.

Clayey banks and cliffs near the sea, rare. E. Kent and Essex; formerly in W. Sussex, but now extinct there. C. and S. Europe.

Represented in England by subsp. **officinale,** which occurs throughout the range of the species.

J. D. Hooker states that the root yields a stimulant resin, but in view of the scarcity of the plant no attempt should be made to confirm this statement on British material.

The only other British umbellifer which could be mistaken for this species is *Foeniculum*, which is easily distinguished by the absence of fibres at the base of the stem and by the absence of bracteoles.

There are about 100 species of *Peucedanum*, widely distributed in Europe, Asia and Africa.

Fruit × 5; fruit t.s. × 5.

64. **Peucedanum palustre** (L.) Moench

Milk-Parsley

An almost glabrous biennial with a watery latex in the young parts. *Stems* up to c. 150 cm, hollow, strongly grooved and angled, often purplish, without fibres at the base. *Lower leaves* 2- to 4-pinnate, lanceolate to ovate in outline; lobes 0.5–2 cm, pinnatifid, lanceolate to ovate in outline, with a finely serrulate cartilaginous margin and acute cartilaginous apex; petiole long, strongly canaliculate above and often puberulent beneath, with a short brown or purple, usually auriculate sheathing base; *upper leaves* smaller, less divided and with an entirely sheathing short petiole. *Umbels* compound, with 15–40 somewhat unequal rays 1.5–5 cm long, papillose-puberulent on the inner side; peduncle longer than the rays, usually scabrid at the top; terminal umbel with hermaphrodite flowers, the lateral umbels with about equal numbers of male and hermaphrodite flowers. *Bracts* 4 or more, linear-lanceolate or sometimes divided, deflexed; bracteoles several, linear-lanceolate, deflexed. *Flowers* white; sepals present, obtuse; outer petals not radiating; styles with enlarged base, forming the stylopodium. *Fruit* 4–5 mm, elliptical, dorsally compressed, smooth: commissure broad; mericarps with rather prominent dorsal ridges and narrowly winged lateral ridges, the wings flat, closely appressed to one another and rather thick; carpophore present; vittae solitary; pedicels longer than the fruit, papillose; styles rather longer than the stylopodium, recurved; stigma capitate. *Cotyledons* abruptly contracted into a petiole. $2n = 22$. Flowering from July to September.

Fens and marshes, local. Mainly in East Anglia, with a few other scattered stations in the southern half of England. Most of Europe, extending eastwards to C. Asia.

According to J. D. Hooker the root abounds in a yellow fetid gum-resin.

This and *Selinum carvifolia* grow in similar habitats but may be easily distinguished. *P. palustre* has hollow, often purplish stems, pinnatifid leaf-lobes and deflexed bracteoles, while *Selinum* has solid, greenish stems, entire or sometimes lobed leaf-lobes and erecto-patent bracteoles.

P. palustre is well known to entomologists as the food plant of the larvae of the Swallow-tail butterfly.

Fruit × 7; fruit t.s. × 7.

65. **Peucedanum ostruthium** (L.) Koch

Masterwort

An almost glabrous perennial. *Stems* up to c. 100 cm, hollow, striate, often purplish below, without fibres at the base. *Lower leaves* 1- to 2-ternate; lobes usually 5–10 cm, lanceolate to ovate, often unequal at the base, irregularly biserrate, the teeth with a long cartilaginous point, the middle lobe often 3-fid, the lateral lobes sometimes 2-fid; petiole long, with a slightly inflated sheathing base; *upper leaves* small, with often pinnatifid lobes and the petiole greatly inflated and entirely sheathing. *Umbels* compound, with usually 30–60 somewhat unequal rays 1–5 cm long, papillose on the adaxial side; peduncle longer than the rays, somewhat puberulent at the top; terminal umbel with hermaphrodite flowers, the lateral with male flowers. *Bracts* absent, rarely 1 or 2; bracteoles few, linear. *Flowers* white; sepals minute; outer petals not radiating; styles with enlarged base, forming the stylopodium. *Fruit* 3.5–5 mm, suborbicular, dorsally compressed, smooth; commissure broad; mericarps with rather prominent dorsal ridges and winged lateral ridges, the wings about as wide as the mericarp, flat, closely appressed to one another, thin; carpophore present; vittae solitary; pedicels mostly longer than the fruit, smooth; styles about twice as long as the stylopodium, recurved, slender; stigma capitate. *Cotyledons* abruptly contracted into a petiole. $2n = 22$. Flowering in July and August.

Formerly cultivated as a pot-herb and for veterinary purposes and naturalized in a number of places, mainly in N. England and in Scotland, where it grows in grassy places and near rivers. Native of the mountains of C. and S. Europe.

The glabrous leaves with large, irregularly biserrate lobes and the umbels with very numerous, slender, papillose rays are distinguishing features of *P. ostruthium*.

Fruit × 7; fruit t.s. × 7.

66. Pastinaca sativa L.

Wild Parsnip

An erect, more or less hairy biennial with a strong smell. *Stems* up to 180 cm, hollow or solid, furrowed and angled, without fibrous remains of petioles at their base. *Leaves* simply pinnate, with (2–)5–11 pairs of lobes, the lobes usually 2–5(–10) cm, occasionally pinnatifid, ovate, cuneate to rounded at the base, coarsely serrate, usually puberulent on both surfaces; petiole of lower leaves inflated near the base, that of upper leaves inflated throughout its length. *Umbels* compound, with 4–11(–17) somewhat unequal, puberulent-papillose rays 1.5–8 cm long; peduncles variable in length; terminal umbel with hermaphrodite flowers towards the outside and male flowers towards the middle, the later lateral umbels with male flowers only. *Bracts* and bracteoles 0–2, caducous. *Flowers* yellow; sepals absent; outer petals not radiating; styles with enlarged base, forming the stylopodium. *Fruit* 5–7 mm, suborbicular, dorsally compressed, smooth; commissure broad; mericarps with slender, low dorsal ridges and rather narrowly winged lateral ridges, the wings flat, closely appressed to one another; carpophore present; vittae solitary, usually rather shorter than the mericarp and tapering at the ends; pedicels as long as or longer than the fruit, puberulent-papillose; styles somewhat longer than the stylopodium, recurved, rather stout; stigma capitate. *Cotyledons* abruptly contracted into a petiole. $2n = 22$. Flowering in July and August.

Common on roadsides and in grassy places south and east of a line from the Severn estuary to the Humber; scattered and often introduced elsewhere. Most of Europe; W. Asia.

Two subspecies can be recognized in Britain:

Subsp. **sativa**: Hairs on stem, petiole, upper surface of leaves and on rays rather sparse, short and straight. Leaf-lobes often narrow and acute. Probably an escape from cultivation in most, if not all, localities.

Subsp. **sylvestris** (Miller) Rouy & Camus: Hairs on stem, petiole, upper surface of leaves and on rays abundant, usually long and flexuous. Leaf-lobes usually broad and obtuse. Commoner than subsp. *sativa* and usually with a preference for base-rich soils. This subspecies is illustrated.

P. sativa may be recognized by the simply pinnate leaves, yellow flowers and strong smell of parsnip.

The genus contains about 15 species and occurs in Europe and temperate Asia.

  Fruit × 5; fruit t.s. × 5.

67. **Heracleum sphondylium** L.

Hogweed, Cow Parsnip, Keck

An erect, usually more or less hispid biennial. *Stems* up to 200(–300) cm, hollow, ridged, often with deflexed but not closely appressed hairs, particularly below. *Leaves* simply pinnate, very rarely 2-pinnate, the larger cauline with usually 5(–9) often pinnatifid lobes; lobes 4–10(–18) cm, very variable in shape, usually rather densely hairy beneath, less so above, crenate to serrate; petiole greatly inflated, sheathing, often purplish, more or less hairy. *Umbels* compound, with usually 10–20 hairy, somewhat unequal rays 2–12 cm long; peduncle longer than rays, more or less puberulent; terminal umbel with hermaphrodite flowers, the lateral umbels with male and hermaphrodite or only male flowers. *Bracts* few or absent; bracteoles usually present, linear, hispidulous, often deflexed. *Flowers* white, greenish-white or rarely pink; sepals very small; outer petals radiating or not; styles with enlarged base, forming the stylopodium. *Fruit* (6–)7–8(–10) mm, obovate to suborbicular, dorsally compressed, smooth or sparsely hairy; commissure broad; mericarps with slender, low, dorsal ridges and rather broadly winged lateral ridges, the wings flat and closely appressed to one another; carpophore present; vittae solitary, about $\frac{1}{2}$–$\frac{3}{4}$ as long as the fruit, up to 0.4 mm wide, widest near the lower end; pedicels as long as or longer than the fruit, puberulent on the adaxial side; styles about twice as long as the stylopodium, divergent or somewhat recurved; stigma capitate. *Cotyledons* abruptly contracted into a petiole. $2n = 22^*$. Flowering from June to September.

Grassy places and open woods, common and generally distributed. Throughout North Temperate Regions.

A very variable species with a large number of subspecies, two of which occur in Britain:

Subsp. **sphondylium**: Petals white or pink, the outer strongly radiating. Mainly in N.W. Europe. The illustration shows this subspecies.

Subsp. **sibiricum** (L) Simonkai: Petals greenish-white, the outer not or scarcely radiating. E. Norfolk, perhaps elsewhere in East Anglia. Mainly in N.E. and E.C. Europe, but also in C. and S.W. France.

H. sphondylium is a very common stout hispid roadside umbellifer, almost always with simply pinnate leaves. It flowers for most of the summer and early autumn and its large, flat or slightly convex umbels are much frequented by beetles and other insects.

There are about 70 species of *Heracleum* in north temperate regions and on mountains in the tropics.

Fruit $\times$ 3.5; fruit t.s. $\times$ 2.5; flower from middle of umbel (left) $\times$ 4; flower from margin of umbel (bottom) $\times$ 4.

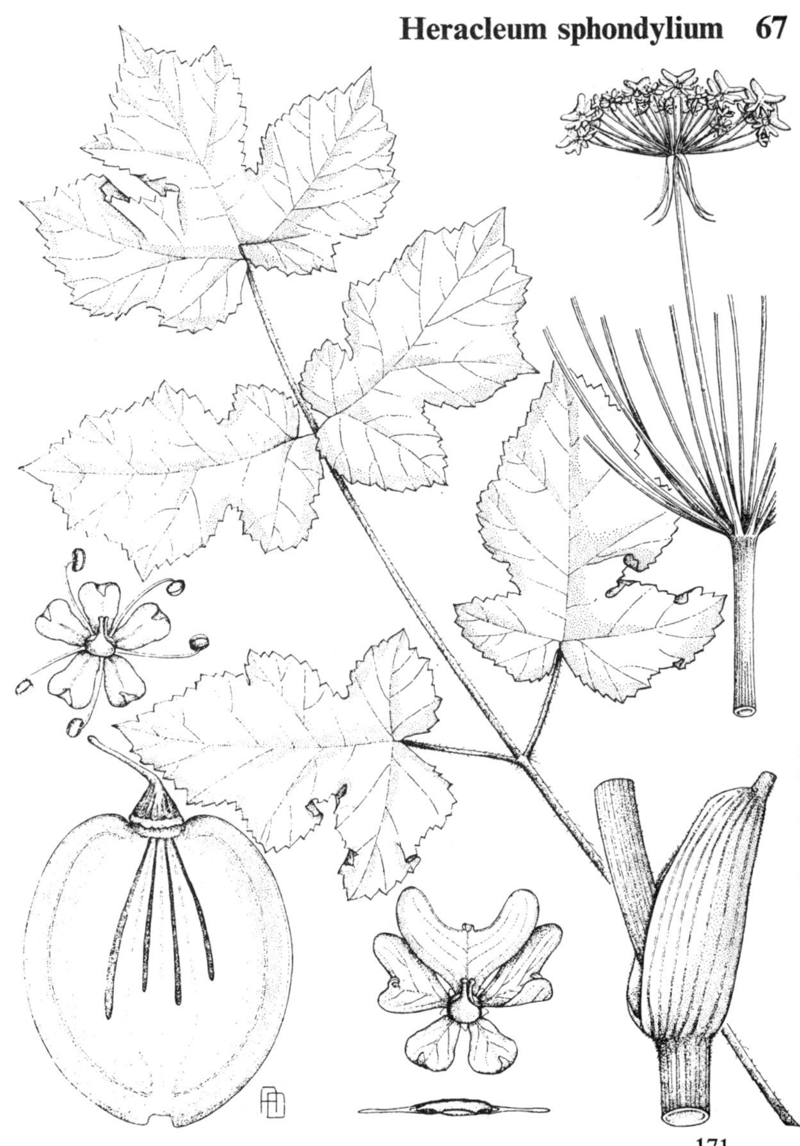

Giant Hogweed

A biennial or monocarpic perennial, sometimes perennating by buds in the axils of the basal leaves. *Stems* up to c. 550 × 5 cm, hollow, ridged, usually purple-spotted, more or less hairy. *Lower leaves* up to c. 250 cm, ternately or pinnately divided to a varying extent, coarsely and irregularly serrate, the lobes and larger teeth long-acuminate, usually puberulent beneath and more or less glabrous above; petiole stout, shortly sheathing at the base; *upper leaves* small, with a greatly inflated sheathing petiole. *Umbels* compound, with 50–150 somewhat unequal, hairy rays 15–30 cm; peduncle as long as to somewhat longer than the rays, more or less hairy; terminal umbel with hermaphrodite flowers, the lateral with mostly male flowers. *Bracts* usually several, linear or ovate and long-acuminate; bracteoles linear. *Flowers* white; sepals triangular; outer petals radiating; styles with enlarged base, forming the stylopodium. *Fruit* 9–11 mm, elliptical, usually glabrous, dorsally compressed; commissure broad; mericarp with slender, low dorsal ridges and rather broadly winged lateral ridges, the wings flat and closely appressed to one another; carpophore present; vittae solitary, c. $\frac{2}{3}$ as long as the fruit, usually c. 1 mm wide at the lower end; pedicels 10–20 mm, hairy; styles c. 3 times as long as the stylopodium, divergent or somewhat recurved; stigma capitate. *Cotyledons* abruptly contracted into a petiole. $2n = 22$. Flowering in June and July.

A native of S.W. Asia, grown for ornament and naturalized on waste ground and near rivers in many scattered localities.

The plant has a strong resinous smell and can cause dermatitis when it is handled in bright sunlight.

Hybrids between *H. mantegazzianum* and *H. sphondylium* have been reported, and one or possibly two somewhat smaller perennial species are sometimes planted and may become naturalized. Their identity and provenance is uncertain.

Fruit × 4; fruit t.s. × 3; flower from middle of umbel (right) × 1; flower from margin of umbel (top centre) × 1.

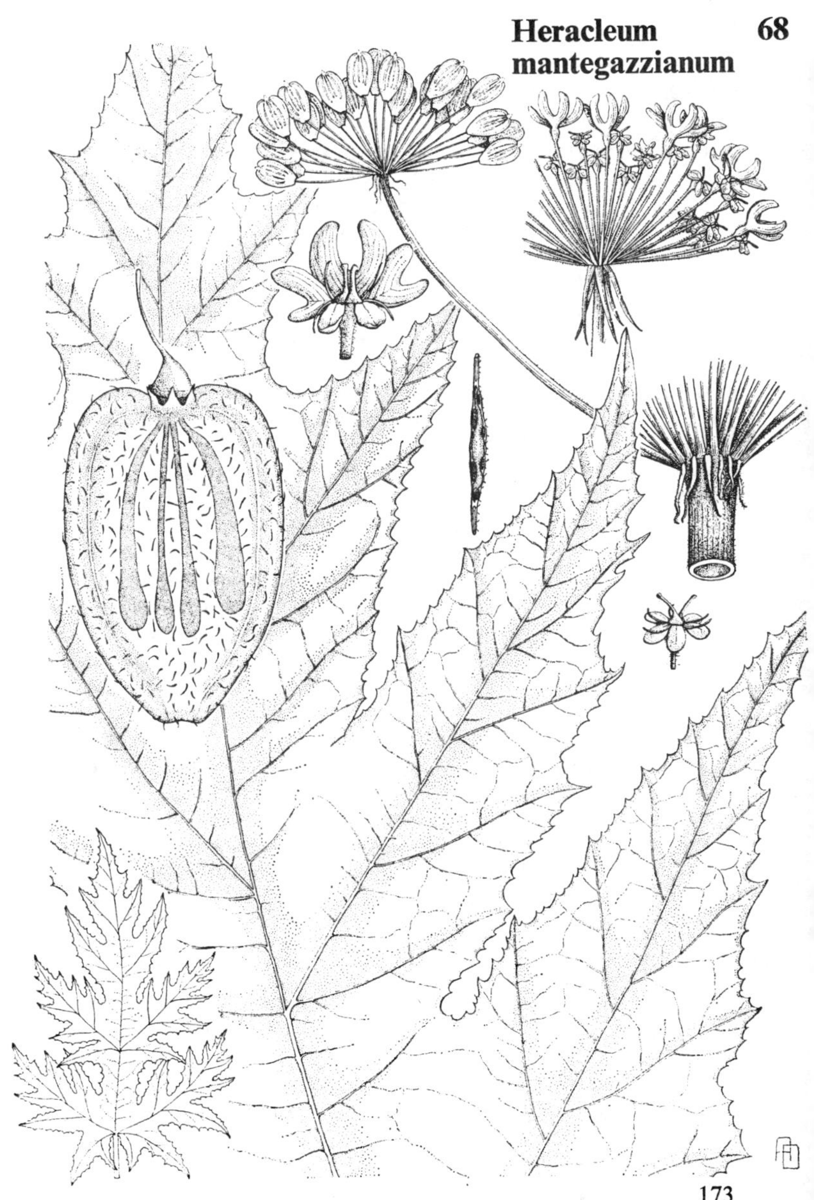

69. **Tordylium maximum** L.

Hartwort

An erect annual or biennial. *Stems* up to 130 cm, hollow to almost solid, ridged, with short, closely appressed, deflexed bristles. Outer basal *leaves* simple, the rest simply pinnate, the lower with usually 3–5 ovate to suborbicular, crenate lobes which are cordate at the base, the cauline leaves with usually 5–11 ovate-lanceolate to linear-lanceolate coarsely serrate lobes which are cuneate at the base and sometimes reduced to the terminal lobe only, all appressed-hispidulous on both surfaces; petiole of basal and lower cauline leaves scarcely sheathing, of upper narrow but sheathing. *Umbels* compound, with (3–)5–15 densely hispidulous rays 0.5–1.5 cm long; peduncle longer than rays, with short, closely appressed, deflexed bristles; umbels with all flowers hermaphrodite. *Bracts* and bracteoles 4–7, linear, hispidulous. *Flowers* white; sepals conspicuous, about half as long as the petals; outer petals radiating; styles with enlarged base, forming the stylopodium. *Fruit* 5–6 mm, elliptical, setose, dorsally compressed; commissure broad; mericarp with slender low ridges and surrounded by a whitish wing with a thin inner part and thick, rounded outer part; carpophore present; vittae solitary, as long as the fruit; pedicels much shorter than the fruit, hispidulous; styles stout, divergent; stigma capitate. *Cotyledons* abruptly contracted into a petiole. $2n = 20$. Flowering in June and July.

Very local in S.E. England in grassy places, mostly along the Thames estuary, where it is perhaps native. S.C. and S. Europe, S.W. Asia. The genus includes 16–20 species in Europe, S.W. Asia and North Africa.

The large sepals and strongly radiating outer petals characterize the plant in flower, while the fruit is unmistakable.

Fruit × 5; fruit t.s. × 4.

70. **Torilis nodosa** (L.) Gaertner

Knotted Hedge-Parsley

A usually procumbent annual. *Stems* up to c. 50 cm, solid, striate, with rather sparse, closely appressed deflexed bristles. *Leaves* 1- to 2-pinnate, the lobes up to c. 3 cm, pinnatifid, lanceolate to ovate in outline, with appressed forward-pointing bristles on both surfaces; petiole of lower leaves slender, with short sheathing base and forward-pointing appressed bristles, the upper leaves subsessile or with a short sheathing petiole. *Umbels* compound, leaf-opposed, with 2-3 short stout rays, which are generally hidden by the flowers or fruit; peduncle usually less than 1 cm, often almost absent, with rather sparse, appressed deflexed bristles; umbels with hermaphrodite flowers only. *Bracts* absent; bracteoles linear, longer than the subsessile flowers, with patent or forward-pointing bristles. *Flowers* pinkish-white; sepals acute, small; petals hispid beneath, the outer not radiating; styles with enlarged base, forming the stylopodium. *Fruit* 2.5–3.5 mm, ovoid, subterete; commissure narrow; mericarp with slender ciliate ridges, the outer with long glochidiate spines and the inner densely tuberculate between the ridges; carpophore present; vittae solitary; pedicels with a ring of hairs at the apex; styles about half as long as the stylopodium, erect or somewhat divergent; stigma capitate. *Cotyledons* tapered gradually at the base, without a distinct petiole. $2n = 22, 24$. Flowering from May to July.

On dry, rather bare banks and in arable fields, local; England, Wales, southern Scotland and the southern half of Ireland. S. and W. Europe, S.W. Asia, North Africa.

Easily recognized by the dense, subsessile leaf-opposed umbels.

T. leptophylla (L.) Reichenb.fil. is like *T. nodosa* in having the umbels mostly apparently lateral and leaf-opposed, but is readily distinguished by the longer, easily visible rays. It is native in S. Europe and is a fairly common casual.

Torilis contains 8 species and occurs in Europe and temperate Asia.

Fruit × 7; fruit t.s. × 5.

71. **Torilis arvensis** (Hudson) Link

Spreading Hedge-Parsley

A usually erect annual. *Stems* 5–40(–100) cm, often freely branched with the branches spreading at a wide angle, solid, terete and often glabrous below, somewhat angled and with sparse to dense closely appressed deflexed bristles above. *Leaves* 1- to 2-pinnate or sometimes 3-lobed, the lobes up to c. 5 cm, coarsely serrate to pinnatifid, lanceolate in outline, with appressed, forward-pointing bristles on both surfaces; petiole slender, with appressed, forward-pointing bristles. *Umbels* compound, with usually 3–5 somewhat unequal rays 0.5–1.5 cm long, with appressed, forward-pointing hairs; peduncle usually longer than the rays, with closely appressed deflexed hairs; umbels with hermaphrodite flowers only. *Bracts* absent or 1; bracteoles several, linear to lanceolate, hispid, about equalling the shortly pedicellate flowers. *Flowers* white or pinkish; sepals acute, persistent; petals hispid beneath, the outer radiating; styles with enlarged base, forming the stylopodium. *Fruit* 4–6 mm, oblong-ovoid, subterete; commissure narrow; both mericarps with slender ciliate ridges and long, rough, tapering spines thickened near the apex and with a slender hook; carpophore present; vittae solitary; pedicels with a ring of hairs at the apex; styles as long to twice as long as the stylopodium, patent or recurved, usually with a few hairs near the base; stigma capitate. *Cotyledons* tapered gradually at the base, without a distinct petiole. $2n = 12$. Flowering in July and August.

Arable fields, mainly in the south-eastern part of England, doubtfully native and very much less frequent now than formerly. W., C. and S. Europe, S.W. Asia.

Represented in Britain by subsp. **arvensis,** which occurs throughout the range of the species.

Mostly easily distinguished from the following, much commoner, species by the usual absence of bracts.

Fruit × 6; fruit t.s. × 8.

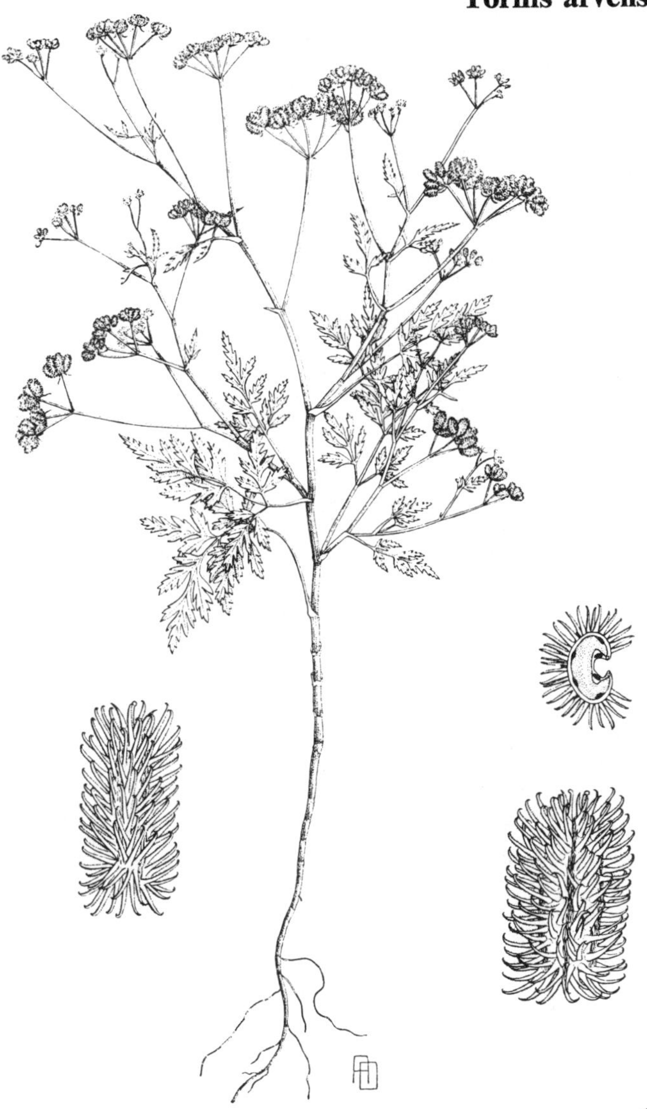

72. **Torilis japonica** (Houtt.) DC.

Upright Hedge-Parsley

An erect annual or rarely biennial. *Stems* up to c. 125 cm, solid, striate, with closely appressed deflexed bristles. *Leaves* 1- to 3-pinnate, the lobes usually 1–3 cm, lanceolate to ovate in outline, pinnatifid or coarsely serrate, with appressed forward-pointing bristles on both surfaces; petiole slender, with mostly deflexed appressed bristles. *Umbels* compound, with (4–)6–10(–13) somewhat unequal rays 0.5–2 cm long, with appressed forward-pointing hairs; peduncle longer than the rays, with closely appressed deflexed hairs; umbels with hermaphrodite flowers only. *Bracts* 4–6(–12), linear, hispid, unequal; bracteoles several, linear to lanceolate, hispid, the longer about equalling the pedicels. *Flowers* pinkish- to purplish-white; sepals conspicuous, acute, persistent; petals hispid beneath, the outer scarcely radiating; styles with enlarged base, forming the stylopodium. *Fruit* 3–3.5 mm, oblong-ovoid, subterete; commissure narrow; both mericarps with slender, almost glabrous ridges and rough, tapering, forward-curved but not hooked spines of varied length; carpophore present; vittae solitary; pedicels with a ring of hairs at the apex; styles about three times as long as the stylopodium, recurved, glabrous; stigma capitate. *Cotyledons* tapered gradually at the base, without a distinct petiole. $2n = 16$. Flowering in July and August.

In hedges and grassy places throughout most of the British Isles. Most of Europe, temperate Asia, Japan, North Africa.

The commonest roadside umbellifer flowering in July, just after *Chaerophyllum temulentum*, in most parts of the country.

Caucalis platycarpos L., Small Bur-Parsley (*C. lappula* Grande), and **Turgenia latifolia** (L.) Hoffm., Great Bur-Parsley (*Caucalis latifolia* L.), formerly occurred as weeds in arable fields but are now very seldom seen. They are both annuals with spiny fruits. *Caucalis* has no bracts (rarely 1–2 small ones) and the outer petals scarcely radiating, while *Turgenia* has (2–)3–5 conspicuous bracts and the outer petals strongly radiating. They are now most likely to be found on waste ground or rubbish-tips. They both have $2n = 20$ and flower in June and July.

Fruit × 10; fruit t.s. × 10.

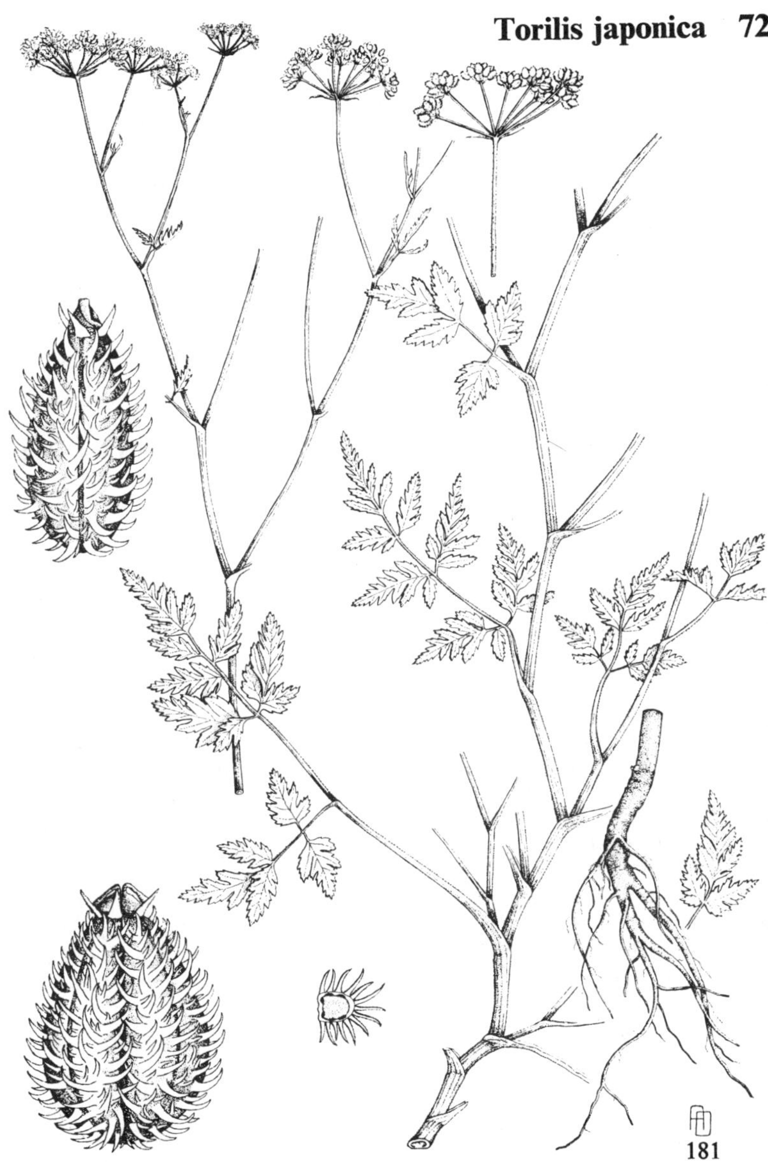

181

73. **Daucus carota** L.

Wild Carrot

Biennial. *Stems* up to c. 100 cm, solid, striate or ridged, with more
or less dense patent to deflexed hairs. *Leaves* (1–)2- to 3-pinnate, the
lobes usually 0.5–3 cm, lanceolate to ovate in outline, pinnatifid or
coarsely serrate, obtuse and mucronate to acuminate, more or less
hairy; petiole of lower leaves slender, of upper leaves sheathing but
not greatly expanded. *Umbels* compound, with usually numerous,
somewhat unequal, more or less hairy rays 1–5 cm; peduncle longer
than the rays, more or less hairy. *Bracts* 7–13, pinnatisect or 3-sect,
with linear lobes; bracteoles usually 7–10, linear-lanceolate, almost
entirely scarious, ciliate, those of the outer partial umbels sometimes
3-fid. *Flowers* white, the central one of the umbel usually dark
purple; sepals triangular; outer petals somewhat radiating. *Fruit*
2–3(–4) mm, ovoid, somewhat compressed dorsally; commissure
wide; mericarps with slender, ciliate primary ridges and broader
secondary ridges bearing a single row of spines; spines flattened,
broad-based, smooth, glochidiate; vittae solitary, under the secondary ridges; pedicels without a ring of hairs at the apex; styles 3–4
times as long as the stylopodium, divergent; stigma capitate.
Cotyledons tapered gradually at the base, without a distinct petiole.
$2n = 18^*$. Flowering in June and July.

A very variable species.

Subsp. **sativus** (Hoffm.) Arcangeli, with a fleshy tap-root, is the
cultivated carrot. Two other subspecies occur in the British Isles:

Subsp. **carota**: Root tough, not fleshy. Leaves thin, hairy, not shiny.
Umbel strongly contracted in fruit, with glabrous or shortly hairy rays.
Common in grassy places, particularly on chalky soils, in the south-
eastern part of England, local elsewhere. Most of Europe, temperate
Asia and N. Africa. The illustration is of this subspecies.

Subsp. **gummifer** Hooker fil.: Root tough, not fleshy. Leaves
rather thick, glabrous or sparsely hairy and glossy above. Umbel not
contracted in fruit, the rays with patent to recurved long hairs. Sea
cliffs and dunes on the Atlantic coasts of Britain, France and N. Spain.

The pinnatisect or 3-sect bracts, the usually dark purple central
flower of the umbel and the spiny fruits make *D. carota* easily
recognizable.

The genus contains about 20 species, which occur in the temperate
parts of the northern hemisphere.

                                          Fruit $\times$ 7; fruit t.s. $\times$ 7.

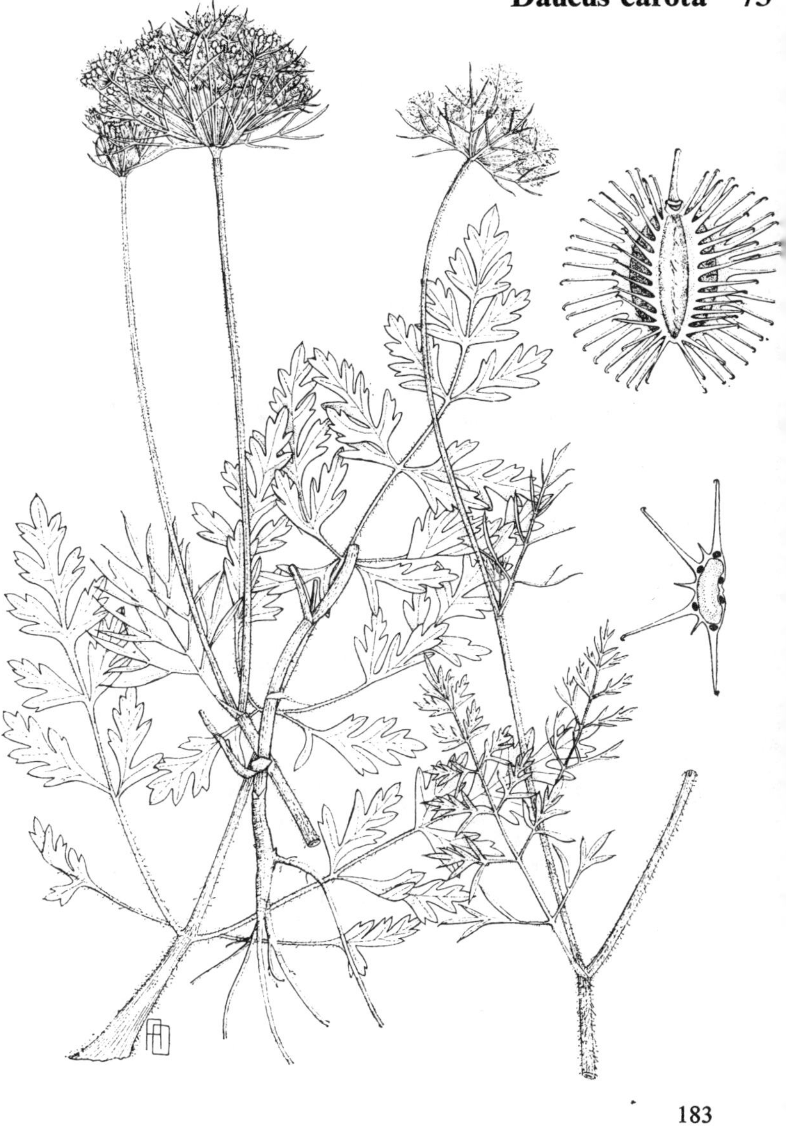

GLOSSARY

accrescent   Becoming larger after flowering.
acuminate   Of a sharply pointed apex with slightly concave sides.
acute   Sharply pointed.
adaxial   Facing the axis or apex.
amplexicaul   Clasping the stem.
annual   Completing its life-cycle within 12 months from germination.
antrorse   Directed towards the apex, forward-pointing.
anther   The part of the stamen containing the pollen-grains.
aristate   With a stiff bristle-like projection at the apex.
auriculate   With small ear-like projections.
beak   Extension of carpel beyond the seed-bearing part.
biennial   Completing its life-cycle in more than one but within two
    years from germination, not flowering in the first year.
bipinnate   See pinnate
bract   A leaf- or scale-like structure subtending the rays of the
    umbel.
bracteole   The leaf- or scale-like structures subtending the flowers
    in a partial umbel.
caducous   Falling early.
calcicole   More frequently found upon or confined to soils con-
    taining free calcium carbonate.
calcifuge   Not normally found upon soils containing free calcium
    carbonate.
canaliculate   With a longitudinal groove.
capitulum   Head of sessile flowers usually surrounded by an
    involucre of bracts.
carpel   One of the units of which the ovary is compressed.
carpophore   The vascular bundle running longitudinally between
    the two carpels and separating from them at maturity.
cartilaginous   Resembling cartilage in consistency.
chromosomes   Small deeply staining bodies found in all nuclei,
    which determine most or all of the inheritable characters of
    organisms. Two similar sets of these are normally present in
    all vegetable cells, the number (*diploid number*, 2n) usually
    being constant for a given sp. The sexual reproductive cells
    normally contain half this number (*haploid number*, n).

Closely related spp. often have the same number, or a multiple of a common *basic number*.

ciliate   With regularly arranged straight hairs projecting from the margin.

commissure   The face by which the two carpels are joined together.

connate   Of organs of the same kind growing together and becoming joined though distinct in origin.

connivent   Of organs with their bases wide apart but their apices approaching one another.

cordate   Heart-shaped.

cotyledons   The first leaves of the seedling, already present within the seed and differing in shape from the later leaves.

crenate   Of a margin with shallow rounded indentations; diminutive *crenulate*.

crispate   Curled (usually of hairs).

cuneate   With a V-shaped base.

cuspidate   Of a sharply pointed apex with strongly concave sides.

cyme   An inflorescence in which the oldest flower is at the apex and which is consequently determinate.

decurrent   Having the base prolonged downwards on the axis, as in leaves where the blade is continued downwards on petiole or rhachis.

deflexed   Bent sharply downwards.

deltate   Shaped like the Greek letter Δ.

dentate   Having teeth projecting more or less at right angles to the margin; diminutive *denticulate*.

dioecious   Having the sexes on different plants.

diploid   See *chromosomes*.

disc   A fleshy, nectar-secreting, more or less flat structure surmounting the ovary.

divaricate   Diverging at a wide angle.

ellipsoid   Of a solid body elliptical in longitudinal section.

endocarp   Inner layer of fruit wall.

falcate   Sickle-shaped.

-fid   Divided in the outer half, as 2-fid etc.

filament   The stalk of the anther, the two together forming the stamen.

fistular   Hollow but closed at the ends.

fruit   The ripe seeds and structure surrounding them, whether fleshy or dry. The 'seeds' of Umbelliferae are strictly half fruits each containing one seed.

fusiform   Spindle-shaped.

glabrous Devoid of hairs.

glaucous Bluish.

glochidiate Barbed at the apex.

herb A plant which is not woody.

hermaphrodite With stamens and ovary in the same flower.

hirsute Clothed with long, not very stiff hairs.

hispid Coarsely and stiffly hairy; diminutive *hispidulous*.

hypocotyl The part of a seedling between the cotyledons and the root.

inflexed Bent inwards.

inflorescence A flower-bearing portion of stem above the last cauline leaf.

laciniate Deeply and irregularly divided into narrow segments.

lanceolate About three times as long as wide, tapering at both ends and broadest below the middle.

latex Milky juice.

linear More or less parallel-sided and flat with a length to breadth ratio of about 12 to 1.

lobes The ultimate distinctly stalked divisions of a leaf.

loculus A chamber of the ovary.

membranous Thin, dry and flexible, not green.

mericarp A 1-seeded portion split off from a fruit containing two seeds at maturity.

monocarpic Fruiting once after one, two or more years and then dying.

mucronate With a short, narrow point (*mucro*) at the apex; diminutive *mucronulate*.

node The point on a stem where a leaf arises.

ob- (in combinations, e.g. obovate) Inverted; an obovate leaf is broadest above the middle, an ovate one below the middle.

oblong More or less parallel-sided and flat with a length to breadth ratio of 2–6 to 1.

obtuse Blunt.

orbicular Rounded, with length and breadth about the same.

ovary The structure enclosing the ovules and consisting of one or more carpels.

ovate The shape of an egg in longitudinal section, with the broadest part below the middle.

ovoid Egg-shaped, with the broadest part below the middle.

ovule A structure containing the egg and developing into a seed after fertilization.

palmate (of a leaf) Consisting of more than 3 leaflets arising from the same point.

palynology   The study of microspores, including pollen-grains.

papilla   A small projection; adj. *papillose*.

partial umbel   A small umbel forming part of a compound umbel.

-partite   Divided into distinct parts.

patent   Diverging from the axis at almost 90°

pedicel   The stalk of a single flower.

peduncle   The stalk of an inflorescence.

peltate   Of a flat organ with its stalk inserted on the under surface, not at the edge.

perennial   Living for more than 2 years and flowering more than once, normally every year.

perfoliate   Of a leaf the base of which completely surrounds the stem.

pericarp   The fruit wall surrounding the seed or seeds.

petal   A member of a whorl of soft, coloured or white structures lying between the sepals and the stamens and alternating with both.

petiole   The stalk of a leaf.

pinnate   Of a leaf composed of more than 3 leaflets arranged along a common stalk or rhachis; *bipinnate* (2-pinnate), of a leaf in which the primary divisions are themselves pinnate. Similarly, 3-pinnate etc.

pinnatifid   Pinnately cut, but with the lobes connected by lamina as well as midrib or rhachis.

polyploid   Having a chromosome number which is a multiple, greater than two, of the basic number of its group (see *chromosomes*).

pruinose   Having a whitish 'bloom'.

pubescent   Shortly and softly hairy; diminutive *puberulent*.

pyrene   A seed contained within the woody endocarp of the fruit, e.g. the stone of a hawthorn.

radiating   With the outer petals of the peripheral flowers larger than the others.

ray   The stalk of a partial umbel.

recurved   Bent backwards in a curve.

rhachis   The axis of a pinnate leaf.

rhizome   An underground stem lasting more than one growing season.

rhombic   Having more or less the shape of a diamond in a pack of playing cards.

scabrid   Rough to the touch; diminutive *scabridulous*.

scarious   Thin, not green, rather stiff and dry.

-sect Divided to the base, midrib or rhachis.

self-incompatible Incapable of self-fertilization.

sepal A member of a whorl of often green, firm structures lying outside the petals and alternating with them.

septum A partition dividing up a hollow structure internally; adj. *septate*.

serrate Toothed like a saw, with the teeth directed towards the apex; diminutive *serrulate*.

sessile Without a stalk.

setose Bristly.

spathulate Paddle-shaped.

simple Not divided into completely separate parts.

stamen One of the male reproductive organs of the plant.

stigma The receptive surface of the gynoecium.

stellate Star-shaped.

stipule A scale-like or leaf-like appendage at the base of the petiole.

stolon A creeping stem of short duration produced by a plant which has a central rosette or erect stem; adj. *stoloniferous*.

striate Marked with long narrow shallow depressions or low ridges.

strigose Covered with straight rigid close-pressed short hairs; diminutive *strigulose*.

style The structure connecting the ovary with the stigma.

stylopodium The prominent nectar-secreting disc and swollen style-base at the apex of the ovary in the Apioideae.

sub- Nearly, not quite; e.g. subglobose.

subulate Awl-shaped, narrow and pointed.

terete Not ridged, grooved or angled.

ternate Of a compound leaf divided into 3 more or less equal parts, which may themselves be similarly divided (2- or 3-ternate).

truncate With a more or less straight transverse apex or base.

umbel An inflorescence in which the pedicels all arise from the top of the main stem; also used of compound umbels, where the peduncles (rays) also arise from the same point.

undulate Wavy in a plane at right angles to the surface.

vein A strand of strengthening or conducting tissue running through a leaf.

vesicular Bladder-like.

villous Shaggy.

vitta An oil canal.

whorl More than two organs of the same kind arising at the same level.

REFERENCES AND A SHORT BIBLIOGRAPHY

BELL, C. R. 1971. Breeding systems and floral biology of the Umbelliferae, or Evidence for specialization in unspecialized flowers.   In Heywood, V. H. (ed.), Biology and Chemistry of the Umbelliferae, *Bot. Journ. Linn. Soc.* 64, Suppl. 1: 93–107.

BENTHAM, G. & HOOKER, J. D. 1867. Umbelliferae. In *Genera Plantarum*, 1: 859–931. London.

CAUWET, A.-M., CERCEAU-LARRIVAL, M.-T., DURRUTY, M. & GUINOCHET, M. 1975. Ombellifères. In Guinochet, M. & Vilmorin, R. de, *Flore de France*, fasc. 2: 382–485.

CAUWET-MARC, A. M. & CARBONNIER, J. (eds.) 1978. Les Ombellifères: Contributions Pluridisciplinaires à la Systématique. Actes du 2e Symposium International sur les Ombellifères. Perpignan.

CERCEAU-LARRIVAL, M.-Th. 1962. Plantules et Pollens d'Ombellifères. *Mém. Mus. Nat.* ser. B, 14: 1–166.

CONSTANCE, L. 1971. History of the classification of the Umbelliferae. In Heywood, V. H. (ed.), Biology and Chemistry of the Umbelliferae, *Bot. Jour. Linn. Soc.* 64, Suppl. 1: 1–11.

DRUDE, C. G. O. 1897–1898. Umbelliferae. In Engler, A. & Prantl, K. (eds.), *Die natürlichen Pflanzenfamilien* 3(8): 63–250. Leipzig.

FROEBE, H. 1971. Inforescence structure and evolution in Umbelliferae. In Heywood, V. H. (ed.), Biology and Chemistry of the Umbelliferae, *Bot. Jour. Linn. Soc.* 64, Suppl. 1: 157–176.

HEDGE, I. C. 1973. Umbelliferae in 1672 and 1972. *Notes Roy. Bot. Gard. Edinb.* 32: 151–158.

HEDGE, I. C. & LAMOND, J. M. 1964. A guide to the Turkish genera of Umbelliferae. *Notes Roy. Bot. Gard. Edinb.* 25: 171–177.

HEGI, G. 1925–1926. Umbelliferae. *Ill. Fl. Mitteleur.* ed. 1, 5(2): 926–1537. München.

HERRNSTADT, I. & HEYN, C. G. 1975. A study of *Cachrys* populations in Israel and its application to generic delimitation. *Bot. Not.* 128: 227–234.

HEYWOOD, V. H. (ed.). 1971. Biology and Chemistry of the Umbelliferae. *Bot. Journ. Linn. Soc.* 64, Suppl. 1.

HOWARTH, S. 1976. *Herbs with everything.* London.

KOMAROV, V. L. et al. (eds.). 1950–1951. *Flora U.R.S.S.* vols. 16–17. Mosqua & Leningrad.

MORISON, R. 1672. *Plantarum umbelliferarum distributio nova.* Oxford.

TUTIN, T. G. 1962. Umbelliferae. In Clapham, A. R., Tutin, T. G. & Warburg, E. F., *Flora of the British Isles,* ed. 2: 495–531. Cambridge.

TUTIN, T. G. (ed.). 1968. Umbelliferae. In Tutin, T. G. et al. (eds.), *Flora Europaea,* 2: 315–375. Cambridge.

WOLFF, H. 1910. Umbelliferae. In Engler, A. (ed.), *Das Pflanzenreich* IV. 228, 43: 1–214; 1913, *op. cit.* 61: 1–305; 1927, *op. cit.* 90: 1–398.

INDEX

References given are to the species number, not page number.
For species of culinary interest not indexed, see pp. 1–415.

BSBI PUBLICATIONS

Handbooks

Each Handbook deals in depth with one or more difficult groups of British and Irish plants.

No. 1   ***Sedges of the British Isles***
A. C. Jermy, A. O. Chater & R. W. David. Revised edition, 1982. 272pp., with descriptions, line drawings and distribution maps for all 73 species. Paperback. ISBN 0 901158 05 4.

No. 2   ***Umbellifers of the British Isles***
T. G. Tutin. 1980. 200pp., with descriptions and line drawings of 73 species. Paperback. ISBN 0 901158 02 X. New edition with distribution maps in preparation.

No. 3   ***Docks and knotweeds of the British Isles***
J. E. Lousley & D. H. Kent. 1981. 208pp., with descriptions and line drawings of about 80 native and alien taxa. Paperback. ISBN 0 901158 04 6. Out of print. New edition with distribution maps in preparation; orders being recorded.

No. 4   ***Willows and poplars of Great Britain and Ireland***
R. D. Meikle. 1984. 200pp., with descriptions and line drawings of 65 species, subspecies, varieties and hybrids. Paperback. ISBN 0 901158 07 0.

No. 5   ***Charophytes of Great Britain and Ireland***
J. A. Moore. 1986. 142pp., with descriptions and line drawings of 39 species and varieties and 17 distribution maps. Paperback. ISBN 0 901158 16 X.

No. 6   ***Crucifers of Great Britain and Ireland***
T. C. G. Rich. 1991. 342pp., with descriptions of 148 taxa (129 with line drawings) and 60 distribution maps. Paperback. ISBN 0 901158 20 8.

No. 7   ***Roses of Great Britain and Ireland***
G. G. Graham & A. L. Primavesi. 1993. 208pp., with descriptions and line drawings of 13 native and nine introduced taxa, descriptions of 76 hybrids, and 33 maps. Paperback. ISBN 0 901158 22 4.

No. 8   ***Pondweeds of Great Britain and Ireland***
C. D. Preston. 1995. 352pp., with descriptions and line drawings of all 50 species and hybrids, most with distribution maps; detailed introductory material and bibliography. Paperback. ISBN 0 901158 24 0.

No. 9   ***Dandelions of Great Britain and Ireland***
A. A. Dudman & A. J. Richards. 1997. 344pp., with descriptions of 235 species, with silhouettes of herbarium specimens, drawings of bud involucres for 139 species, and 178 distribution maps. Paperback. ISBN 0 901158 25 9.

Available from the official agents for BSBI publications, F. & M. Perring, Green Acre, Wood Lane, Oundle, Peterborough PE8 4JQ (Tel. 01832 273388. Fax 01832 274568).